ONE EARTH

OUR CHANGING GLOBAL ENVIRONMENT

ONE FUTURE

by
Cheryl Simon Silver
with
Ruth S. DeFries
for the
National Academy of Sciences

NATIONAL ACADEMY PRESS
Washington, D.C. 1990

National Academy Press • 2101 Constitution Avenue, N.W. • Washington, D.C. 20418

NOTICE: The project that is the subject of this report was approved by the Governing Board of the National Research Council, whose members are drawn from the councils of the National Academy of Sciences, the National Academy of Engineering, and the Institute of Medicine.

This report has been reviewed by a group other than the authors according to procedures approved by a Report Review Committee consisting of members of the National Academy of Sciences, the National Academy of Engineering, and the Institute of Medicine.

Support for this project was provided by The Business Roundtable, the Arthur L. Day Fund of the National Academy of Sciences, the Geraldine R. Dodge Foundation, the Department of Energy, the Environmental Protection Agency, the National Aeronautics and Space Administration, the National Oceanic and Atmospheric Administration, the National Science Foundation, the Tinker Foundation, and the U.S. Committee for Man and the Biosphere. The Rockefeller Foundation contributed to the support of the teleconferencing.

This book is printed on acid-free recycled paper. 

Library of Congress Cataloging-in-Publication Data

Silver, Cheryl Simon.
One earth, one future : our changing global environment / by Cheryl Simon Silver with Ruth S. DeFries for the National Academy of Sciences.
p. cm.
ISBN 0-309-04141-4 (case); ISBN 0-309-04632-7 (paper)
1. Man—Influence on nature. 2. Pollution. 3. Global warming. 4. Environmental policy. I. DeFries, Ruth S. II. National Academy of Sciences (U.S.) III. Title.
GF75.S55 1990 90-5939
363.7—dc20 CIP

Printed in the United States of America

First Printing, November 1990
Second Printing, July 1992
Third Printing, November 1992
Fourth Printing, January 1994
Fifth Printing, January 1996

Cover photograph:

ROUSSEAU, Henri
The Dream. 1910.
Oil on canvas, 6′8 1/2″ × 9′9 1/2″.
Collection, The Museum of Modern Art,
York New York. Gift of Nelson A. Rockefeller.
Photograph © The Museum of Modern Art, New York

Preface

Human activities are transforming the global environment, and these global changes have many faces: ozone depletion, tropical deforestation, acid deposition, and increased atmospheric concentrations of gases that trap heat and may warm the global climate. For many of these troubling transformations, data and analyses are fragmentary, scientific understanding is incomplete, and long-term implications are unknown. Yet, even against a continuing background of uncertainty, it is abundantly clear that human activities—burning fossil fuels, emitting pollutants from industry, and clearing forests that are the habitats for plant and animal species, for example—now match or even surpass natural processes as agents of change in the planetary environment. Understanding the nature and possible consequences of global change is an urgent challenge to the natural sciences, social sciences, and engineering, and to the world community of nations and their citizens.

Global environmental change is interwoven with a complex web of social, economic, political, and scientific implications. Recent natural fluctuations in weather and climate, while not necessarily attributable to climate change due to human activities, nevertheless illustrate the magnitude and broad scope of environmental impacts on our intricately intertwined global economy. Widespread droughts in the early 1970s set the stage for major worldwide swings in grain prices. Persistent droughts in Africa have caused unspeakable suffering for millions of people. Damage caused as pollutants travel across national boundaries and result in acid deposition has created major political tensions in North America.

The diverse faces of global environmental change are linked both scientifically and politically. Scientifically, the ability to predict future changes in the environment requires an understanding of the physical, chemical, biological, and social processes that govern the earth, and of the interaction of these processes throughout the earth system. Politically, policy options to address these problems highlight the need for coordinated international policies relating to energy, technology, land use, and economic development.

Thus difficult policy decisions must be made on the basis of judgments between dimly perceived future risks and possible economic or other consequences that may be more immediate. While these decisions must be based on the best information that science has to offer, scientists are no better qualified than other individuals to hammer out these difficult judgments. It is important that the public also become informed and involved in making these choices and shaping the necessary policy decisions.

It was with a realization of the indispensable role of an informed public that the 1989 Forum on Global Change and Our Common Future was conceived. The forum's goal was to promote and enhance understanding and dialogue on a broad range of issues related to global environmental change and the dynamic interactions among the physical, chemical, biological, and social systems that make the earth's environment so uniquely hospitable to life. The presentations and discussions during the eventful three days of the forum provided the stimulus and the basis for this book.

The forum is but one of many activities related to these

complex issues. For example, in the fall of 1988 the National Academy of Sciences, National Academy of Engineering, and Institute of Medicine prepared recommendations on global environmental change for then President-elect George Bush. That document, which may be found in Appendix D, is the current position of the Academies and the Institute in this area. In addition, at the request of Congress, the U.S. Environmental Protection Agency commissioned a study on policy implications of greenhouse warming by the Committee on Science, Engineering, and Public Policy of the councils of the two Academies and the Institute. The report of that panel, which is expected to be available by the end of 1990, will be the next major statement of the Academies and Institute in this area. Within this context, this book is intended as a contribution to the active global dialogue that will shape the future of our species and our planet.

FRANK PRESS
President
National Academy of Sciences

Acknowledgments

Many people contributed to the development of this book. First and foremost, the Forum on Global Change and Our Common Future, on which this book is based, was originally conceived by Harold Mooney of Stanford University, chairman of the National Research Council's Committee on Global Change (see Appendix C). Tom Malone, of St. Joseph College, and many others were instrumental in developing and organizing the forum. John Perry, of the National Research Council, was particularly crucial in the development of both the forum and this book and provided helpful guidance along the way.

We gratefully acknowledge the participation of the speakers at the forum (see Appendix B), whose papers provided the basis for this book. Informal comments on technical aspects of each of the chapters of this book were provided by the following experts in the relevant fields, and we thank them for their time and thoughtful critiques:

INTRODUCTION William Clark, Harvard University; Robert Kates, Brown University

A NEW SCIENCE OF THE EARTH Kevin Burke, Lunar and Planetary Institute; S. Ichtiaque Rasool, National Aeronautics and Space Administration; Myron F. Uman, National Research Council

LESSONS FROM THE GEOLOGIC PAST Kevin Burke, Lunar and Planetary Institute; John Kutzbach, University of Wisconsin; S. Ichtiaque Rasool, National Aeronautics and Space Administration; Myron F. Uman, National Research Council

SYSTEM INTERACTIONS: ATMOSPHERE, OCEANS, LAND, AND HUMANS Kevin Burke, Lunar and Planetary Institute; S. Ichtiaque Rasool, National Aeronautics and Space Administration; Myron F. Uman, National Research Council

HUMANITY: AN AGENT OF GLOBAL ENVIRONMENTAL CHANGE Jesse Ausubel, Rockefeller University; William Clark, Harvard University; Robert Kates, Brown University; B. L. Turner II, Clark University

GLOBAL WARMING Robert Dickinson, National Center for Atmospheric Research; Jerry Mahlman, National Oceanic and Atmospheric Administration; Stephen H. Schneider, National Center for Atmospheric Research

FOOD, WATER, AND CHANGING CLIMATE Norman Rosenberg, Resources for the Future; Cynthia Rosenzweig, Columbia University

COASTLINES AND RISING SEAS James Broadus, Woods Hole Oceanographic Institution; Mark Meier, University of Colorado

THE OZONE LAYER AND ULTRAVIOLET RADIATION Dan Albritton, National Oceanic and Atmospheric Administration; F. Sherwood Rowland, University of California at Irvine; Susan Solomon, National Oceanic and Atmospheric Administration

VANISHING FORESTS AND VANISHING SPECIES Judith Gradwohl, Smithsonian Institution; Robert Peters, World Wildlife Fund; Jeffrey Richey, University of Washington

LAKES, FORESTS, AND ACID DEPOSITION James Galloway, University of Virginia; Mark Haag, Department of Justice; Arthur Johnson, University of Pennsylvania; David Schindler, University of Alberta, Edmonton, Canada

Roseanne Price, of the National Research Council, edited the draft and provided many helpful comments on the content.

We gratefully acknowledge the contributions of all of these people, without whom this book would not have been possible.

CHERYL SIMON SILVER
RUTH S. DeFRIES

Contents

THE FACES OF GLOBAL ENVIRONMENTAL CHANGE

AFTERWORD

APPENDIXES

1

Introduction

As the human race prepares to venture into a new century, conversations and news reports are peppered with references to our "fragile and endangered planet." This phrase almost certainly exaggerates the case. The earth is 5 billion years old, and over the eons it has endured bombardment by meteors, abrupt shifts in its magnetic fields, dramatic realignment of its land masses, and the advance and retreat of massive ice mountains that reshaped its surface. Life, too, has proved resilient: In the more than 3.5 billion years since the first forms of life emerged, biological species have come and gone, but life has persisted without interruption. In fact, no matter what we humans do, it is unlikely that we could suppress the powerful physical and chemical forces that drive the earth system.

Although we cannot completely disrupt the earth system, we do affect it significantly as we use energy and emit pollutants in our quest to provide food, shelter, and a host of other products for the world's growing population. We release chemicals that gnaw holes in the ozone shield that protects us from harmful ultraviolet radiation, and we burn fuels that emit heat-trapping gases that build up in the atmosphere. Our expanding numbers overtax the agricultural potential of the land. Tropical forests that are home for millions of biological species are cleared for agriculture, grazing, and logging. Raw materials are drawn from the earth to stoke the engines of the growing world economy, and we treat the atmosphere, land, and waters as receptacles for the wastes generated as we consume energy and goods in our everyday lives.

Scientific evidence and theory indicate that as a result of such activities, the global environment is undergoing profound changes. In essence, we are conducting an uncontrolled experiment with the planet.

The changes facing the planet today are distinguished from previous changes by the scale and pace with which they are occurring or are likely to occur. Over the geologic past, conditions in the atmosphere, ocean, and biosphere have for the most part followed natural cycles. Now, human activities are a significant force driving changes in the global environment.

This book is intended to present briefly the current state of scientific knowledge about the changes under way in the global environment. Its organization, inspiration, and, to a substantial extent, its content derive from the Forum on Global Change and Our Common Future, which took place in May 1989 in Washington, D.C. (see Appendixes A, B, and C), and which was jointly sponsored by the National Academy of Sciences, the Smithsonian Institution, the American Association for the Advancement of Science, and Sigma Xi, the scientific research society. Speakers at the forum placed the current state of the global environment firmly in the context of the earth system, in which human activity is increasingly viewed as a counterpart to the integrated workings of the atmosphere, oceans, land masses, and biosphere.

Humans have long affected their local environment, but it is only in this century, particularly in the last 50 years, that the scope of this influence has expanded to a global scale. Now it is clear that with ceaseless repetition, even seemingly innocuous actions such as driving a car or cutting down a tree can influence the physical and chemical systems that govern the earth. In the case of the acid deposition that alters soil chemistry and affects the natural chemical balance of many of the world's waterways and lakes, the effects of pollutant emissions in one part of the world can be felt half a continent away. The chlorofluorocarbons (CFCs) produced and emitted by industrialized countries rise to the stratosphere, where, like other pollutants, they disperse without regard to political or geographical boundaries. When forest land is cleared and used for grazing cattle or growing crops, plant and animal communities are disrupted (or, for species with limited habitat, eliminated).

Our activities have already caused changes in the composition of the earth's atmosphere, most notably increased atmospheric concentrations of trace gases such as carbon dioxide, methane, nitrous oxide, and chlorofluorocarbons. From the beginning of earth's history, certain of these trace gases have had an important role in the principle that is known as the "greenhouse effect." In a fashion crudely analogous to the function of the glass panels of a greenhouse, the gases keep some of the earth's heat from escaping through the atmosphere and thus have the potential to warm the earth's surface temperature. Atmospheric concentrations of carbon dioxide—after water vapor the most plentiful greenhouse gas—have increased significantly, mostly because humans have burned vast quantities of fossil fuels in the past 100 years, releasing carbon dioxide in the process. As forests, which absorb carbon dioxide as they grow, are cleared and the felled trees and brush are burned or decay, the massive amounts of carbon a forest ordinarily stores are released to the atmosphere.

If rapid warming of the earth's surface from changes in the composition of the atmosphere occurs—as scientists warn is within the realm of possibility—other global and regional changes could result. Sea level could rise as ice caps melt and the ocean expands from the extra heat, and agricultural belts will shift. Forests and other ecosystems may be torn apart as both plant and animal groups respond to vastly altered temperature and hydrological conditions, and as various kinds of plants adjust differently to increased carbon dioxide levels. Rapid changes in climate could cause regional food shortages, create waves of environmental refugees, and threaten the security of other countries as the effects of the turmoil ripple through the world economy. This is not to say that all changes will be bad. Some northern high-latitude regions could witness increased agricultural yields and benefit from a warmer climate.

With the physical, chemical, and biological changes in the earth system that humans are causing, more surprises like the discovery of the antarctic ozone hole cannot be ruled out. Atmospheric chemists point out that despite years of study, no one predicted that a hole would form each year in the protective ozone layer high in the atmosphere over Antarctica. Unexpected events of comparable or greater gravity are likely to continue to appear even as scientific understanding evolves.

"Nature seems to be running a fever. We are the flu," observed William Ruckelshaus, former head of the U.S. Environmental Protection Agency, at the Forum on Global Change and Our Common Future in May 1989. "Our goal is not so much to manage the planet earth as to make ourselves less like a pathogen and more like those helpful bacteria that dwell in our own guts."

The transformation of the global environment is driven by ever-greater numbers of people, increasing economic development, and its attendant increase in industrial activity and consumption of energy by humans. Of these factors, population growth is the most easily quantified. Since 1900, the number of people has more than tripled. In 1987, the 5 billionth member entered the human family. Our numbers are increasing today by about 90 million per year, and, according to United Nations (U.N.) projections, are expected to reach 10 billion or more by the end of the coming century. Ninety-five percent of these people will be born in the poor countries of today's developing world.

While population has increased, so have standards of living for many of the earth's people, consumption of fossil fuels, and expansion of the world economy. These changes have allowed astonishing improvements in human welfare, but at a cost. Unfortunately, many of the processes that produce gains degrade the environment and deplete the ecological capital—the soils, forests, species, fisheries, and water resources—on which humanity relies.

People in the richer, developed countries, with only one quarter of the world's population, consume most of the world's energy. They command about 80 percent of the world's wealth, use most of its natural resources, and generate the most waste. Most of the greenhouse gases and chemicals that are changing the composition of the atmosphere—and thus contributing to the projected climate change and to other changes such as acid deposition—have been emitted by industrialized nations in the Northern Hemisphere.

People in the developing countries, with three quarters of the world's population, have less than one quarter of the wealth. But the millions of poor people in the developing world also contribute to resource depletion and environmental stress. The poor and hungry are often compelled to destroy their environment—by cutting down forests and depleting soils—in order to survive.

In the developing world, improved standards of living can break the cycle of rapid population growth and the environmental hazards it engenders. The experience of country after country has shown that economic development, when paired with better opportunities for employment and education, eventually leads to lower birth rates. The catch, as Ruckelshaus noted at the forum, is that, "If the four-fifths of humanity now in developing nations attempts to create wealth using the methods of the past, at some point the result will be unacceptable world ecological damage." Developing countries now account for about one quarter of all greenhouse emissions. If the same roads to prosperity are followed that were taken in the past by developed nations, this share could double by the middle of the coming century.

The prospect of unchecked global environmental change thus raises troubling questions about equity that world leaders can no longer ignore. Since sacrifice and gain will be unequally distributed, what is the responsibility of the "have" nations that created many of the problems to the "have not" nations that seek to develop in order to meet the needs of their people? Is it equitable to expect the poorer nations in their pursuit of improved standards of living to hold to stringent environmental controls that the developed nations did not have to contend with in their periods of rapid industrial growth? And what is the responsibility of people now living to future generations?

The existing disparities between the rich countries of the North and the poor countries of the South and the lack of capital and infrastructure in poor countries to cope with rapid changes in environmental conditions are prompting John Holdren, of the University of California at Berkeley, Jessica Mathews, of the World Resources Institute, and many other prominent political analysts, world leaders, and politicians to reconsider the nature of possible future threats to world peace and security. For decades the main threat to security has been widely associated with the potential for conflict between East and West, with the nuclear arms race, and with the U.S./Soviet tensions centered in Central Europe. In the late 1980s, as epochal changes in Central Europe forced the world to rapidly revise its perceptions of the nuclear threat, the potential risks arising from changes in the global environment began to claim attention. "The potential explosion of tensions deriving from global environmental change is clearly going to be aggravated by the widespread perception that the biggest burdens will fall on the developing countries of the South and that the principal culprits in generating these problems, through both action and inaction, are in the developed countries of the North," said Holdren at the forum. International dialogue that is currently under way represents the opportunity for nations—developed and developing alike—to overcome these tensions and attack together common concerns about the global environment.

The future of the global environment and the consequences of the changes in store assumed new currency, at least within political circles, in 1987 when the World Commission on Environment and Development issued its path-breaking report, *Our Common Future*. (The World Commission is also known as the Brundtland Commission after its chairperson, Gro Harlem Brundtland, former Prime Minister of Norway [see Afterword, p. 147].) The commission, created in 1983 by the U.N. General Assembly, was a special independent organization that included representatives from 22 countries, a majority of them developing countries. Over a 3-year period, the members tackled the question tacitly posed by the General Assembly: Is it possible to meet the needs of the 5 billion people alive today without compromising the ability of future, ever more populous generations to meet their own needs?

After exhaustive study and analysis, the commission answered with a heavily qualified "yes." The grim but very real prospect that the future of humanity depends on an environment ever more polluted, degraded, and devoid of the ecological resources required to ease poverty and hardship could be averted if processes of economic and social development are transformed. The commission invoked the concept of sustainable development to describe a means by which economic and social progress could be achieved without compromising the integrity of the environment.

Sustainable development is not a new concept, but its prominence in *Our Common Future* has stimulated governments, policymakers, economists, and moralists throughout the world to reconsider its implications. Of particular interest is the Brundtland Commission's central theme: the integration of environment and economy. As Brundtland pointed out in her keynote address to the forum, "Only growth can eliminate poverty and create the capacity to solve environmental problems. But growth cannot be based on overexploitation of the resources of developing countries. It must be managed to enhance the resource base on which these countries all depend."

If this new doctrine is to govern future approaches to growth and development, painful choices will be required in how we exploit resources, direct investments, develop technologies, and organize our institutions. The wealthy nations must recognize that their continued prosperity depends in part on maintaining the earth's ability to supply food and other resources and that this in turn requires increased prosperity and security in the developing world. *Our Common Future* heightened awareness of some widely discussed, and thought-provoking, arguments. One suggestion, for example, is that it may be time to price goods to reflect the environmental costs of their production and use. Gasoline, for instance, might be priced to reflect the costs of the damage that burning the fuel causes through pollution and increased risk of climate change. Likewise, the price of fertilizer might reflect the cost of cleaning up the water supplies it pollutes. In addition, *Our Common Future* suggests that it may be appropriate to restructure institutions to reflect environmental priorities. This would mean that environmental aspects of policy would be considered at the same time as questions of economics, trade, energy, and agriculture, and by the same national and international institutions. Environmental agencies would be given more power to redirect policies that now lead to environmental degradation. Economic, trade, and other government agencies heavily endowed with money and power would be mandated to develop policies that encourage sustainable development and would be responsible to their governments for the environmental consequences of their policies and budget allocations.

The Brundtland Commission emphasized that if the pace of global change is to be checked, developing countries are likely to need fresh infusions of financial support in the 1990s to pay for efforts to reduce rates of population growth, to restore and maintain natural resources, and to adopt modern technologies that are less polluting than the ones already outdated in the industrialized world. *Our Common Future,* as well as many political and economic analysts, points out that the question of increased aid cannot be taken seriously until the debt situation is resolved. Debt remains an urgent problem facing developing countries, particularly in Latin America and Africa. The World Bank reports that in 1988 the 17 most indebted nations paid developed nations and multilateral agencies $31.1 billion more than they received in aid.

As governments, industrialists, scientists, and the public

consider the implications of suggestions such as those above, it is increasingly apparent that the decisions that will make these adjustments possible will require tighter connections between science and policy.

The earth has already been committed to major environmental change in the years ahead. The elevated concentrations of greenhouse gases already emitted through human activity will persist for many centuries, no matter what we do. The chlorofluorocarbons in the atmosphere today will continue to deplete the ozone shield for centuries. Complex tropical forest ecosystems once cleared can regenerate slowly at best. The magnitude and rate of change will depend on whether societies decide to act to slow the atmospheric buildup of carbon dioxide and other trace gases, reverse deforestation, and cut pollutant emissions. Steps required to slow the pace of change, and for adapting to it if need be, will be costly, but so may be the costs of inaction. As climatologist Stephen H. Schneider, of the National Center for Atmospheric Research, cautions, if society chooses to wait another decade or more, "we face a higher risk that we will have to adapt to a larger amount of climate change than if actions to slow down the buildup of greenhouse gases were pursued more vigorously today."

In the face of much scientific, social, and political uncertainty, encouraging signs abound. Demographers report, for instance, that birth rates in many countries are declining. Forty-nine nations have ratified the Montreal Protocol on Substances That Deplete the Ozone Layer, which calls for a 50 percent reduction in chlorofluorocarbon production from 1986 levels by 1999 and provides a model for cooperation that spans national boundaries and interests. The amount of energy and raw materials required to produce a given amount of goods is decreasing in some countries, including the United States.

Scientists can provide information on which to base decisions, but whether to act to slow the pace of environmental change is a social judgment, not a scientific one. Among scientists there is a broad consensus that even in the face of incomplete or conflicting data, measures to ease adaptation to the potential changes, or even to reduce the rate of changes already under way, are possible and advisable. Steps such as improving energy efficiency and developing alternatives to fossil fuels not only would slow the rate of warming but also could buy crucial time to study climate change and assess the impacts. The presidents of the National Academy of Sciences, the National Academy of Engineering, and the Institute of Medicine advised then President-elect George Bush in 1988 that, "While global environmental change cannot be stopped, the pace of change can be slowed. We cannot buy absolute security against environmental risk, however much we are willing to pay, but we may be able to reduce environmental damage and risk markedly by prudent policy actions" (see Appendix D).

In the end, the outcome will depend on political will. The social, political, and economic changes would be enormous. Yet government leaders are increasingly attuned to the gravity of the hazards facing society as environmental degradation proceeds. They are talking at the highest levels, and many are struggling to translate acceptance of a set of values, such as the need to protect the environment, into action. In the United States, our environmental statutes are among the most stringent in the world, and the people repeatedly express their desire for increased environmental protection. Yet this country is the largest emitter of greenhouse gases, and environmental pollution continues to be a major problem.

The chapters that follow provide the backdrop of scientific information on which the political decisions ultimately will rely. The complexities of earth system science, the lessons derived from the earth's history, and the modern forces driving changes in the global environment are explored in the first part of this book, "The Earth as a System." The second part, "The Faces of Global Environmental Change," describes some of the transformations projected or under way. These include the prospect of a warmer global climate, potential changes to the world's food production systems and water supplies as climate changes, and the likelihood that sea level will rise as the climate warms and how this may affect the world coastal zones and their inhabitants. Subsequent chapters examine other human-induced changes to the global environment, including the threat to the earth's ozone layer, deforestation and dwindling genetic diversity, and how acid deposition affects our forests, lakes, and waterways. (Appendix A contains a selected list of additional reading on global environmental change.)

The environmental challenges described are difficult ones for a world already grappling with other costly and more obvious problems of economics, security, and public health. But as the following chapters demonstrate, global environmental change may well be the most pressing international issue of the coming century.

THE EARTH AS A SYSTEM

have allowed us, for instance, to venture into space, to develop the scientific understanding necessary to address the challenges we face in protecting the global environment on our own planet? To understand how human-induced changes—global warming, depletion of the protective ozone layer, acid rain, deforestation, and possibly other changes that have not yet been detected—affect and are affected by the earth system, scientists are studying the interactions among processes in the atmosphere, oceans, and land surfaces, and the plants and animals that inhabit them.

In some ways, this new push to understand the earth is a natural outcome of those first glimpses from the moon, two decades back. The quest to understand how the earth works may not match the excitement of man's footprint on the age-old lunar dust or the thrill of a manned trip to Mars. But what this quest lacks in glamor it makes up in importance for the future of the earth's environment: In one of the broadest scientific inquiries in human history, physical and social scientists are drawing on every resource of technology and intellect to advance understanding of both the natural variability of the earth's processes and the effects of human activities on them.

This new approach to the study of our planet is referred to as earth system science. Its practitioners strive to understand how the world works on a global scale by describing how its parts and their interactions evolved, how they function today, and how they may be expected to function in both the near and distant future. In this light, the earth system is seen as a set of interacting subsystems characterized by processes that vary on spatial scales from millimeters to the circumference of the earth, and on time scales from seconds to billions of years. It has become ever more clear that despite wide separations in distance or time, many processes are connected, and that a change in one component can propagate through the entire system. A 1988 report of the Earth System Sciences Committee to the NASA Advisory Council noted, for example, that "volcanic activity occurs widely along intersections of the earth's crustal plates and is driven by mantle convection on long time scales; yet the effects of eruptions are felt locally within hours or days and then, over larger areas, for months or years because of deposition of dust and gases in the atmosphere."

The NASA report explains that a science of the earth system must aim toward understanding processes governing global change over five broad time scales. Processes operating over the longest time scale, millions or billions of years, encompass the evolution of solid-earth structures and include the internal core and mantle processes that generate the earth's magnetic field. As the ages pass, crustal movements rearrange the continents, oceans open and close, and mountains erode. The oceans and atmosphere formed, their chemical compositions were determined, and life evolved within this long time frame.

Over the time scale of hundreds of thousands of years to millions of years, the earth system witnesses oscillations between ice ages and interglacial periods, development of soils, and shifts in the distribution of biological species, largely in response to cyclical changes in the earth's orbit around the sun.

Decades and centuries, the span of a few human generations, are the time scale over which the oceans, atmosphere, and biota interact to form the physical climate system. These systems are linked by the flow of moisture over the globe in the form of vapor, liquid water, and ice, and they change in response to processes and interactions that occur over much shorter periods ranging from seasons to hours. In this time frame, the earth's biosphere responds to and influences cycles that move key substances such as carbon, nitrogen, phosphorus, and sulfur through the global environment.

Within the fourth time frame, days to seasons, the earth responds to weather, changes in ocean currents, growth and melting of the polar ice caps and sea ice, surface runoff and erosion, and the annual cycles of plant growth and decay.

Finally, each day sees a cycle of heating and cooling, growth and decay, that moves heat, water, and a host of substances among land, air, oceans, and biota. Earthquakes and volcanic eruptions occur suddenly on this shortest time scale in response to adjustments occurring within the solid earth over much longer periods.

Human influence is superimposed on the natural earth system processes operating over these time scales. Human civilization is characterized by modification of the environment—beginning with fire and then agriculture—but until fairly recently, we did not profoundly alter the planet as a whole. Over the past few centuries, however, the sheer expansion in the number of the earth's human inhabitants and the growth in our technological ability to modify the landscape and exploit the earth's bounty of minerals, water, and fossil fuel have profoundly changed the entire earth system. The extent and consequences of these changes are only beginning to be understood.

With all that is known and yet to be learned, how do we synthesize the vast body of knowledge necessary to describe the interactive system that is our earth? It is not enough to simply enumerate processes that are important. Participants in the effort to develop an earth system science have devised a schematic model of the earth system—a working hypothesis of how the parts of the system work together—atmospheric and ocean circulation and dynamics, atmospheric chemistry, terrestrial ecosystems, and the global hydrologic cycle. All of these parts of the system interface continually with human activities and with changes in natural inputs from the sun, from volcanoes, and from other natural causes.

Although processes operating on all time scales influence the earth system, for this conceptual model it is the middle time scale—decades to centuries—that is most relevant to the urgent inquiry into global environmental change. Within individual scientific disciplines, the most advanced models developed for use on this time scale focus on the physics and dynamics of the atmosphere. Models of ocean dynamics and atmospheric chemistry are fairly well developed. The least developed models are those describing terrestrial ecosystems and marine biogeochemical systems, which are difficult to predict and subtle in their nature.

Francis Bretherton, director of the Space Science and Engineering Center at the University of Wisconsin at Madison, led the committee that developed the NASA report on earth system science. He explains that schematic models of this sort offer a vehicle by which scientists of different backgrounds can share, in a useful way, the knowledge that has been acquired so laboriously by the work of the world's scientists. Models also indicate which aspects of the earth system may be the most important ones to measure and help scientists test whether their understanding of how the system works is correct.

Although the global environmental changes discussed in this book are partly due to the by-products of technologies developed during and since the Industrial Revolution, it is our technological prowess that enables scientists to measure and observe the changes and processes under way and engineers to develop sophisticated technologies that reduce the burden on the environment. Scientists are confident that within the next two decades many answers will emerge as data are acquired, cross-referenced, and interpreted. As Bretherton cautions, however, "Our vulnerability to error is greatest not from the things that we include in the model, but from prophesies we leave out entirely."

3

Lessons from the Geologic Past

Conspicuous though we now are, *Homo sapiens* have emerged only within the last few hundreds of thousands of years, a flash of an instant in the planet's 5-billion-year history. The results of our actions are merely superimposed on natural variations in the earth's network of oceanic, atmospheric, and biological systems.

As scientists try to predict the future global environment, they constantly confront an enormous obstacle: incomplete understanding of how physical, chemical, and biological processes affect each other and shape the planet today. What is known about the past, however, demonstrates that climate and the fortunes of earth's inhabitants have been intertwined since life began on earth, and that relatively small changes can have large and unexpected consequences.

Michael McElroy, chairman of the Department of Earth and Planetary Sciences at Harvard University, notes that if earth scientists could work in the laboratory, "We would have hypotheses, we would do experiments, we would manipulate the experiment and we would learn about processes by the conventional iteration of theory and experiment. But as geophysicists or geochemists, we do not have that luxury. We cannot manipulate the earth. Our laboratory is the past. If we imaginatively attempt to understand the changes that took place, in the long term we will have our best chance to predict and to guide our future." In particular, several phases in earth's history shed light on what may be in store with future changes in climate.

FORMATION OF THE OZONE SHIELD

Our lessons from the past begin with a development that occurred more than 1 billion years ago. Early aquatic organisms, blue-green algae, began to use energy from the sun to split molecules of water and carbon dioxide and then recombine them into organic compounds and oxygen—the process known as photosynthesis. Oxygen was used up as organic carbon was converted to carbon dioxide, but not all of it. In a fateful development, oxygen—poisonous to organisms then—began to accumulate in the atmosphere, touching off a massive ecological disaster for primordial organisms. As oxygen built up, the carbon dioxide content began to drop. High in the atmosphere, some molecules of oxygen (O_2) were split as they absorbed energy from ultraviolet rays and formed single atoms of oxygen. When these recombined with oxygen, ozone (O_3) molecules formed, which are very effective absorbers of ultraviolet rays emitted by the sun. The ozone formed a protective shield around the earth, eliminating the threat of irradiation by ultraviolet light. With this development, the land was fit for more complex life. The organisms that first began to cast off the oxygen did not survive the switch to an oxygenated atmosphere, but some others did, and made the pivotal transition from water to land.

PANGAEA

A paroxysmal change that occurred about 300 million years ago can help us understand more about our climate today. At that time, when the age of dinosaurs was just beginning, the movement of the earth's crustal plates caused the two major

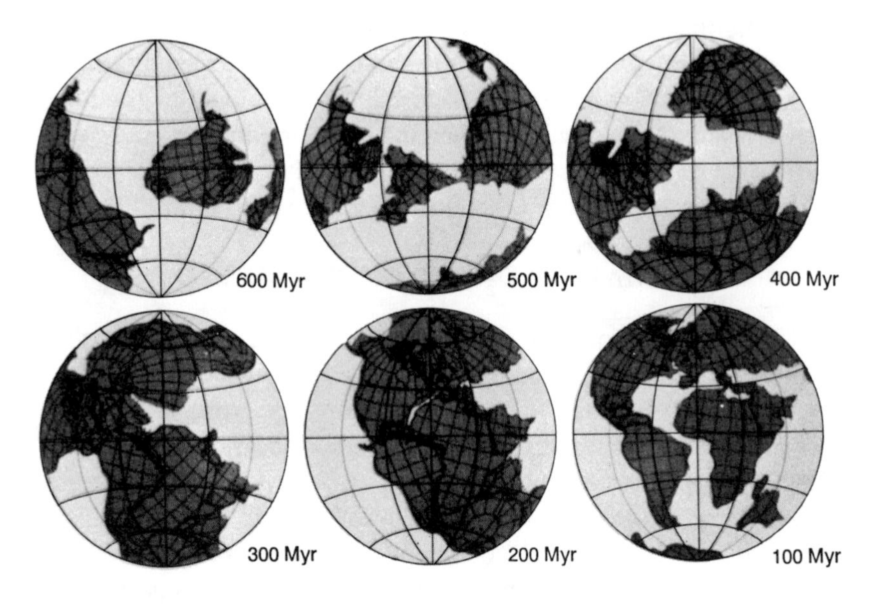

Locations of continents for various times during the past several hundred million years. (Adapted from N. Calder. 1983. *Timescale: An Atlas of the Fourth Dimension*. Viking Press, New York.)

continents at the time—Laurasia in the north and Gondwanaland in the south—to mass together for one unique, relatively brief period of less than 100 million years into a megacontinent called Pangaea, or "all lands."

What was it like on Pangaea? The earth in the age of Pangaea, of course, was very different from the earth today, and from the earth in the preceding and following geologic times, when the continents were not so closely joined. Nevertheless, scientists use this ancient natural experiment to help understand how the distribution of lands and oceans affects climate.

John E. Kutzbach and Robert G. Gallimore, both of the Center for Climatic Research at the University of Wisconsin at Madison, are using a general circulation climate model to calculate the climate of Pangaea. Results from their model, along with evidence from the fossil and geologic record, support earlier speculations that the megacontinent was beset by large-scale monsoon conditions in both summer and winter and that these seasons were typified, respectively, by extreme heat and cold. The continental interior was hot and dry, with monthly average temperatures in summer of 35°C, or well over 100°F. (Since these were average temperatures, many days were probably as warm as almost 50°C, or 120°F.) It was relatively humid in the polar regions and along the coasts of the vast, continental-scale embayment called the Tethys Ocean, where monsoon winds were strong; the tropics, except along the coasts, were dry. These computer simulations of the climate are supported by geologic evidence of the location of plant and animal fossils and mineral deposits; the degree of agreement between model and geologic data is an indication of scientists' growing understanding of how the climate system works. Although the Pangaean world bore little semblance to ours, the global average temperature was only about 5°C (or 9°F) higher than at present.

Before Pangaea formed, the diversity and abundance of life on our planet rose, but as the continents converged into a single mass, the greatest extinction of all time occurred. By some estimates, more than half of all families and three quarters of all species became extinct. Scientists believe that perhaps this great dying between the Permian and Triassic periods was somehow related to the marked changes in climate that accompanied the development of Pangaea. One possibility is that most organisms could not adapt to the extreme fluctuations in temperature and moisture between summer and winter. Today, only Siberia and northern Canada experience as wide a range of seasonal variation. Other possible explanations abound: Did the Pangaean world have too few unique habitats? Did some sort of catastrophe occur? Did changes in the ocean currents, temperature, or salinity disrupt the global climate or the chemical and biological balances? Or did the atmospheric concentrations of oxygen and carbon dioxide perhaps create conditions that the plants and organisms living then simply could not tolerate?

THE ICE AGES

Another significant and telling event in the earth's history was the long slide of climate from warm to cold, beginning about 100 million years ago, when the climate was still much warmer than present and the level of carbon dioxide may have been 10 times greater than it is today. What natural processes could have produced such a high level of carbon dioxide in the earth's atmosphere? It is possible that widespread volcanism had infused the atmosphere with carbon dioxide. The continents dispersed as Pangaea broke apart, and volcanism may have been much more active than today along the mid-ocean ridges where the seafloor was forming and on the continental margins where the seafloor was being subducted. As continents drifted toward their current locations, the seafloor spread apart at a slower rate. Many researchers believe that, consequently, there was less volcanic activity and related carbon dioxide emissions, which led to a cooler climate.

This cooling trend that began almost 100 million years ago, coupled with continued movements of continental plates, led to the growth of huge ice sheets on Antarctica and Greenland. Coincidentally, the North American plate buckled and the Rocky Mountains began to rise. Halfway around the world, the Indian continent collided with the Eurasian continent, giving rise to the Tibetan Plateau, a process still under way. This phase of mountain building may have contributed to conditions conducive to further glaciation of North America and Eurasia in the past 1 million to 2 million years.

To earth scientists, ice ages are in many ways the flip side of the warming that may be in store for us now. In the past decade, great progress has been made toward understanding the cause of the glacial cycles during the most recent 1 million to 2 million years of earth history. While researchers have attempted to explain the causes of ice ages for more than a century, recently developed scientific tools are yielding major new findings.

Some of the most important new information comes from cores carefully extracted by drilling deep into the ice caps of

Greenland and Antarctica. By analyzing the gas in bubbles trapped in the ice, scientists have learned that the atmosphere during the ice ages was quite different from that of today and that the concentrations of greenhouse gases and dust in the atmosphere have undergone wide fluctuations. While several ice cores have been drilled in ice caps around the world, one more than 2 kilometers (1.2 miles) long is particularly valuable because it covers more than an entire glacial cycle. This core, recovered from a drillhole at Vostok in Antarctica, contains the record of 160,000 years of climate history, from the present warm period, or "interglacial," through the most recent 100,000-year-long ice age, through a previous warm period and back into an even earlier ice age.

The Vostok ice core is being analyzed by Claude Lorius and colleagues at the Laboratory of Glaciology and Geophysics of the Environment in St. Martin d'Hères and at the Laboratory of Isotopic Geochemistry in Gif sur Yvette, both in France. While researchers in recent years had already learned that carbon dioxide levels during the most recent ice age were lower than they are during today's interglacial, the French group reports an even stronger relationship between this greenhouse gas and temperature: Atmospheric greenhouse gases and climate generally shift in lockstep throughout the glacial cycle. As the earth moves into an interglacial period, for instance, temperatures rise, and so do concentrations of carbon dioxide. During the deepest part of the ice age, temperatures plummet, and so does carbon dioxide, to perhaps 60 percent of that during the interglacial periods. But researchers do not yet know which is cause and which is effect.

Ice cores are far from the only tool providing insights into the earth's climatic history. Researchers glean clues from other sources such as fossilized pollen grains, annual growth rings of trees, records of changing sea level in coral reefs, and even fossilized middens, the junk piles left by packrats. Cores of sediment extracted from the floor of the deep sea are especially useful because their chemical composition and the warm- or cold-water fossils they contain reflect changes in ocean temperature and the volume of the polar ice caps.

Information from these diverse sources adds up to a picture of growing complexity but increasing clarity. Scientists have established that ice ages are almost certainly triggered by relatively small changes in the amount of sunlight reaching the earth at different latitudes and seasons. These small changes in sunlight, only a few percent, are caused by three orbital effects: slight changes in the earth's elliptical orbit around the sun from nearly circular to more elliptical, over a cycle of about 100,000 years; shifts in the degree at which the earth's axis is tilted, over a cycle of about 40,000 years; and wobbling of the axis itself, over a cycle of about 20,000 years. When the orbital conditions result in less sunlight in summer, the climate cools, ice may gradually accumulate into mountains of ice over 2 miles tall, and, because water is transferred to the ice caps, sea level may drop by several hundred feet. When the orbital conditions yield more sunlight in summer, the climate warms, the ice melts, and sea level rises.

While orbital changes may trigger the glacial cycles, the shifts in sunlight are not great enough in themselves to force climate change of ice age magnitude. Oceans, with their vast capacity for storing heat and carbon, also may play a critical role in causing climate change. Lorius and colleagues suggest that two kinds of ocean fluctuations—"deep changes possibly driven by sea level and surface changes driven by atmospheric circulation"—drive the carbon dioxide variations between glacial and nonglacial times.

Not only physics but also chemistry and biology may be catalysts in the climate cycle. Plankton and other photosynthetic microorganisms living in the ocean may help regulate the world's climate. These microbes absorb carbon dioxide in the process of photosynthesis. According to one scenario, the plankton may flourish, or "bloom," as polar ice caps grow and nutrients in the water become more concentrated or as ocean currents change, bringing nutrient-rich bottom waters to the surface. The explosion in plankton concentrations would mean more photosynthesis. Carbon dioxide levels in the ocean would drop, and consequently more would be pulled from the atmosphere, cooling the earth.

Plankton may also be involved in another mechanism, along with the clouds that form over the ocean and at any given time cover 30 percent of the world. Researchers studying both arctic and antarctic ice say that the concentration of sulfate particles varies with temperature—as temperatures drop, sulfate concentrations rise. One possible reason is that some kinds of plankton excrete a sulfur compound called dimethylsulfide (DMS). When dimethylsulfide diffuses from the ocean to the atmosphere, it oxidizes into sulfate particles, which act as condensation nuclei for water droplets that form marine stratus clouds over the open ocean. As the plankton bloom, the number of nuclei increases. With more nuclei, more incoming solar radiation is reflected back to space by the clouds, lowering the water's surface temperature and cooling the earth.

THE CURRENT WARM PHASE

The geologic record shows that the latest act of the glacial drama opened as the most recent glaciation began to wind down about 18,000 years ago. As has been the pattern, the cold period lasted about 100,000 years; the present balmy climate is a brief warm spell in a typically icy cycle. This most recent switch from a glacial to a warm phase is of special interest to scientists grappling to make sense of the complexities of modern climate because the amount of carbon dioxide that has accumulated in the atmosphere from when the melting began to the present is roughly equal to the amount of greenhouse gases projected to build up from the present to about the middle of the next century.

To determine what happened when the most recent glaciation ended and why, a number of academic institutions have pooled their efforts through the Cooperative Holocene Mapping Project (COHMAP). As Kutzbach explains, they use both geologic data and general circulation climate models to identify

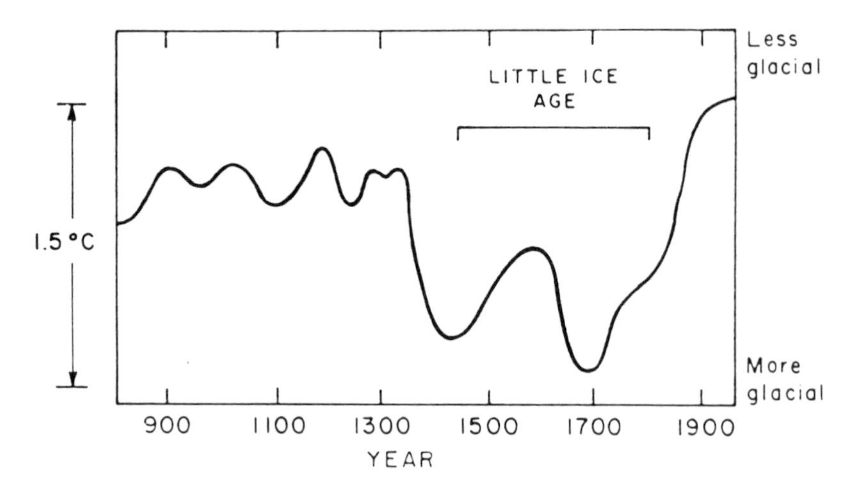

Estimates of the changes in temperature in Europe over the past 1000 years. (Reprinted, by permission, from J. Imbrie and K. P. Imbrie. 1986. *Ice Ages: Solving the Mystery*. Harvard University Press, Cambridge, Mass. Copyright © 1986 by John Imbrie and Katherine Palmer Imbrie.)

and evaluate the causes and mechanisms of this most recent change from glacial to the current interglacial. Although the exact sequence of events is unclear, COHMAP results suggest that the orbitally caused increase of summer sunlight, the rise in carbon dioxide, and the melting of glacial ice all began about 18,000 years ago. The global warming trend over the past 18,000 years has been about 5°C (or 9°F). In high latitudes, such as Canada, the warming since 18,000 years ago has been much greater—10°, 20°, even 30°C. Arctic permafrost and sea ice receded and sea level rose. Spruce forests migrated from the central United States to southern Canada, and lake beds dried up in California and Nevada. These and other environmental changes influenced human communities throughout the world. This most recent global warming of about 5°C occurred over a period of many millennia. In contrast, the projected future warming from human-produced greenhouse gases may occur within a century or so, in other words, perhaps 10 to 50 times faster.

Results from COHMAP also show that climate changes in

the mid-latitudes and the tropics likewise seem to be linked to small changes in the earth's orbit. As radiation in summer increased, the land became hotter, creating greater contrast between ocean and land temperatures. This produced strong summer monsoons from 12,000 to 6,000 years ago in the Northern Hemisphere tropics and subtropics, and drought in the interiors of North America and central Asia. Thus, at a time when the American desert basins were dry, lakes flooded much of North Africa that is today covered by the shifting sands of the Sahara Desert. These lakes formed in shallow depressions of the desert floor. In parts of North Africa and the Middle East, climatic conditions became more favorable for the agricultural revolution then under way.

Agreement between model results and the geologic record of this relatively recent global change reassures scientists that they are now beginning to understand what happens in the climate system.

The climate has also varied in recent centuries and decades. Although these changes were not as dramatic as those of earlier times, nor as large as those expected in the next century, their beginnings and endings are accurately known. From this, scientists know that climate can change abruptly and that the changes can be large enough to have regional impacts. The golden age of the Anasazi Indians on Mesa Verde in the southwestern United States, for instance, may have been cut short by overpopulation and overuse of land, coupled with the persistent drought that began suddenly in the late thirteenth century. This drought began about the time that Europe was gripped by a cold snap, known as the Little Ice Age, that persisted until the nineteenth century. In the 1930s dry conditions led to the North American Dust Bowl. Whether these variations in climate were caused by changes in the amount of sunlight or in the frequency of volcanic eruptions or by subtle internal oscillations of atmosphere and ocean is not known.

As evident from these examples of natural climate variability, it is difficult to recognize the initial phases of human-caused climate change. Nonetheless, it is clear that throughout the history of life on earth, the fortunes of earth's inhabitants have been inextricably tied to variations in climate. To be sure, photosynthesizing algae helped create and maintain the conditions that allowed life to persist without pause for more than 3.5 billion years. We, on the other hand, have produced conditions that could push the earth to the brink of climate change at a rate unprecedented in the planet's history.

It is highly probable that life will survive. After all, life has survived all the past changes of climate. But it may not be life as we know it now. Will plant and animal communities respond quickly enough to the projected environmental change, or will the uneven pace of adjustment literally tear communities apart? Will humans be able to adapt as planetary conditions change? The answer may lie in the planet's past, and in understanding the complex, interdependent components that make up the earth system.

4

System Interactions: Atmosphere, Oceans, Land, and Humans

Over the past several decades, scientists' understanding of the complexities of the earth system has evolved to the point where they now recognize that the components of the system—the atmosphere, oceans, land, and associated living beings including humans—are inextricably intertwined. A change in one part of the earth system has repercussions for other parts—often in ways that are neither obvious nor immediately apparent. It is beyond the human ken, however, to study the whole, multidimensional system at once. As the following sections attest, the effort to understand the dynamics driving change in the global environment is designed along the academic lines that essentially define classical disciplines. In fact, though, researchers are ever-aware that the various sciences of the atmosphere, ocean, land, and water are connected in countless ways. The intricacies of the earth system range from the obvious links between currents in the ocean and atmosphere, to the all-encompassing global cycles of carbon and water, to the subtle, distant effect of clearing a tract of tropical forest on the amount of carbon in the atmosphere.

While each major component of the earth system holds its

mysteries, the effect of human activity on the system can be the greatest wildcard of all. For the first time, the social sciences are assuming substantial weight in the study of the earth system as researchers and policymakers struggle to discern how humankind, this relatively recent, terribly powerful feature of the earth, affects the age-old forces that also dictate our planet's future.

ATMOSPHERE

Many of the earth's inhabitants live far from the oceans; concerns about tropical forests may seem remote to farmers on the American Plains, or to women gathering firewood in the Himalayas. But the atmosphere touches each of us.

The atmosphere, a gaseous envelope that surrounds the earth, is the engine of the physical climate system. When radiation from the sun enters the atmosphere, some is reflected back upward by clouds and dust, and some continues on to the land surface. Of that radiation that strikes the surface, some is absorbed by the earth, but some is reflected back to space by ice, snow, water, and other reflective surfaces. In addition, infrared radiation is emitted by the earth. A portion of this energy gets trapped by certain atmospheric gases whose particular chemistries do not allow the outgoing, longer-wave infrared radiation to escape. Instead, this bounces back to the earth, raising the surface temperature. This phenomenon, which has operated throughout earth's history, is well known as the greenhouse effect.

Without the atmosphere and the greenhouse effect, the earth's surface would be frozen, and life would not be possible. At the other end of the spectrum, the atmosphere on Venus is so dense with carbon dioxide and the greenhouse effect is so intense that the planet's surface is everywhere as hot as cooling lava on earth.

The composition of the atmosphere determines the earth's ability to maintain a balance between the energy coming in and the energy released. The main gases in dry air are nitrogen (79

(79 percent), oxygen (20 percent), and argon (1 percent). Water vapor, present in variable concentrations up to a few percent, is the major gas responsible for the greenhouse effect on earth. Other greenhouse gases are present in trace amounts, usually measured in parts per billion (ppb). The trace gas that has recently received the most attention is carbon dioxide, which currently constitutes 0.034 percent, or 344 parts per million, of the atmosphere.

In addition to carbon dioxide, other trace gases—two chlorofluorocarbons (CFC-11 and CFC-12, which also destroy the protective ozone layer that shields us from harmful ultraviolet radiation), methane, nitrous oxide, and tropospheric ozone—are efficient at absorbing infrared radiation emitted by the earth. They are of special interest now because their concentrations in the atmosphere are rising. As they do, less radiation escapes from the surface into space, and the earth's temperature rises.

The future of the earth's climate and, perhaps, its inhabitants, depends on how much concentrations of carbon dioxide and other trace gases are likely to rise. Carbon dioxide poses the single greatest threat because it is the most abundant of these gases. It occurs naturally in the atmosphere and is cycled through nearly all living organisms. Animals, including humans, exhale it as a waste product, whereas plants "breathe" it, using the carbon to make the carbohydrates they require in the processes known as photosynthesis.

Analysis of air bubbles trapped in glacial ice and contemporary measurements reveal that carbon dioxide concentrations have increased by nearly 25 percent since the eighteenth century, when industrialization began. The main cause is the combustion of fossil fuels, which produces compounds that also contribute to problems such as local air pollution and acid deposition. During combustion, carbon is oxidized to carbon dioxide and released to the atmosphere. The destruction of forests for settlements or cultivation contributes to this rise also. When land is cleared, the trees either decompose or are burned, and the carbon stored in the plant material is released to the atmosphere.

We have accurate records of modern carbon dioxide levels

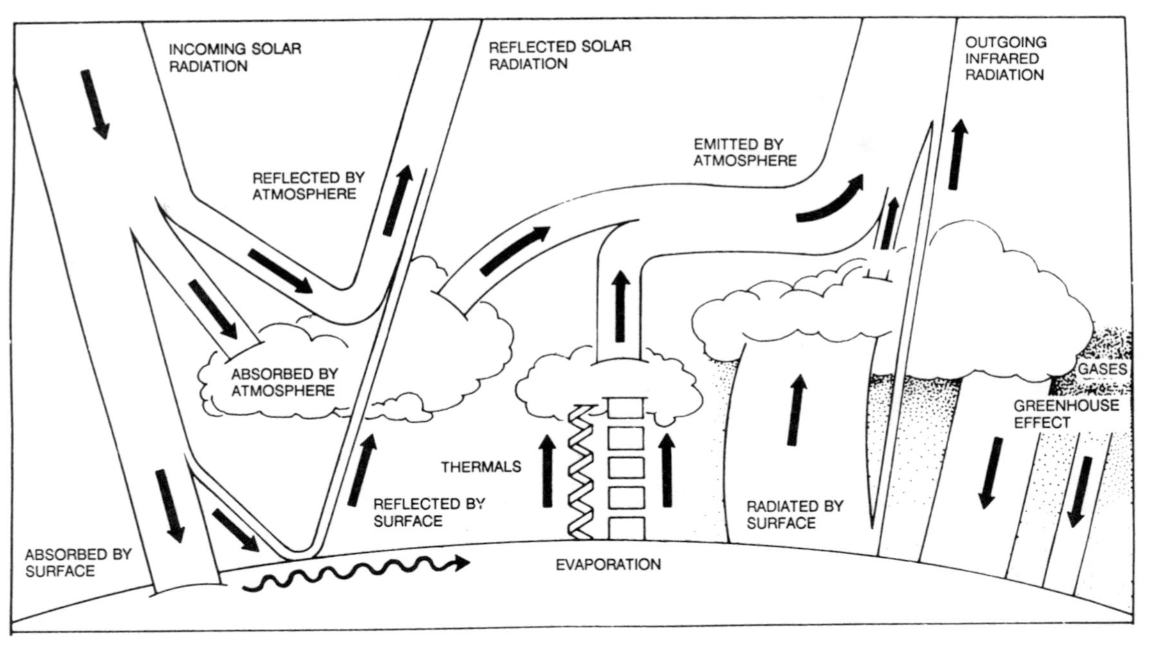

Schematic illustration of the earth's radiation and energy balances. The greenhouse effect is well established. It arises because the earth's atmosphere tends to trap heat near the surface. Carbon dioxide, water vapor, and other trace greenhouse gases are relatively transparent to the visible and near-infrared wavelengths that carry most of the energy of sunlight, but they absorb more efficiently the longer, infrared wavelengths emitted by the earth. Hence an increase in the atmospheric concentration of greenhouse gases tends to warm the surface by downward reradiation of infrared, as shown. (Adapted, with permission, from S. H. Schneider. 1987. "Climate Modeling," *Scientific American*, vol. 256, p. 78.)

since 1958, when Charles D. Keeling, of Scripps Institution of Oceanography in La Jolla, California, began measuring concentrations of atmospheric gases from a research station high on Mauna Loa in Hawaii. His records cover a relatively brief interval, but are treasured by scientists: They clearly show that carbon dioxide is increasing in the global atmosphere, and they also show a striking sawtooth pattern that reflects the entire biosphere of the Northern Hemisphere "breathing in" as plants grow in the warm months and "exhaling" when they are dormant. From studies of glacial ice samples, scientists know that the level of carbon dioxide during ice ages was about 200 parts per million. In between glacial periods, when the earth was warm, it was about 280 parts per million. Today we are at 350 parts per million and climbing.

The other trace greenhouse gases—methane, nitrous oxide, chlorofluorocarbons, and ozone—absorb infrared radiation much more effectively than carbon dioxide does, but they are present in much smaller quantities. Their combined effect may well cause half of the global warming projected for the next century.

The atmosphere's methane content is particularly worrisome because it is rising at a much faster rate than even carbon dioxide. Systematic measurements of methane concentrations did not begin until the late 1960s. During the 1980s, levels of this gas rose sharply, at a rate of about 1.1 percent per year. Studies of ice cores show that the methane increase over the centuries parallels the swelling of human population, a logical connection because methane is produced through the rumination of increasing numbers of cattle and through rice paddy cultivation, which is also increasing. Like carbon dioxide, methane concentrations in the atmosphere vary with the glacial cycle. During the ice ages, methane was present in the atmosphere at roughly 300 parts per billion. During interglacial periods, the atmospheric levels doubled to perhaps 600 parts per billion. Now we are at 1800 parts per billion and climbing. The sources for this rise include melting of tundra permafrost, biomass burning, leaks

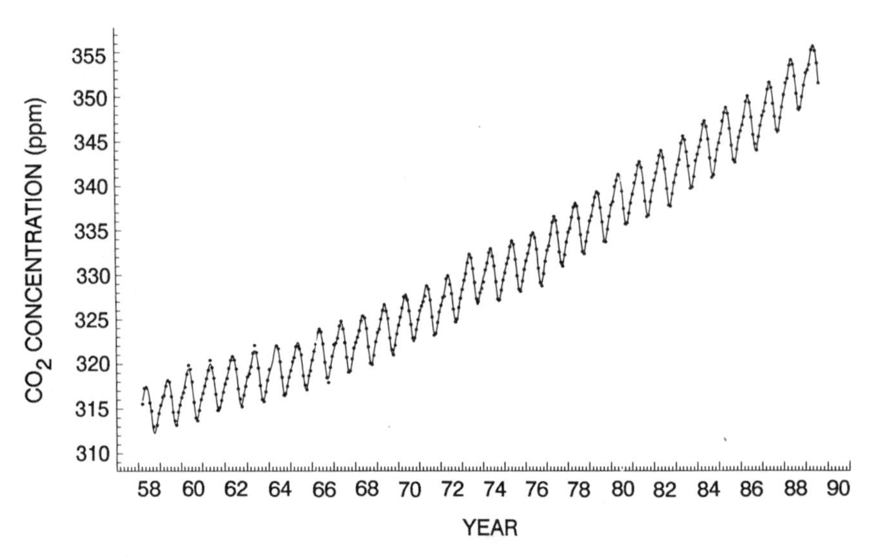

Concentration of atmospheric carbon dioxide in parts per million of dry air (ppm) versus time for the years 1958 to 1989 at Mauna Loa Observatory, Hawaii. The dots indicate monthly average concentration. (Reprinted, by permission, from C. D. Keeling et al. 1989. "A Three Dimensional Model of Atmospheric CO_2 Transport Based on Observed Winds: Observational Data and Preliminary Analysis," Appendix A, in *Aspects of Climate Variability in the Pacific and the Western Americas*, Geophysical Monograph, vol. 55, Nov. Copyright © 1989 by the American Geophysical Union.)

in natural gas pipelines, and emissions from rice paddies and cattle, but they are far from quantified.

Atmospheric scientists try to decipher the workings of the physical climate system by constructing what are known as general circulation models. These computer models use mathematical equations to express the basic physical principles that govern the global atmosphere and then use actual data to test whether the models adequately simulate reality.

The general circulation models, as they now exist, simulate the physical climate and geographical features on a very coarse scale. A country the size of Japan, for example, does not appear on the computer-generated maps. Vast numbers of calculations and large amounts of computer time and money would be required to refine the scale.

It is not only the coarse scale of the general circulation models that is proving problematic in using the models to answer questions about climate change. The current models do not incorporate other components of the earth system that are known to exert strong influences on the physical climate. Scientists are attempting to incorporate the dynamics of the ocean, and its enormous abilities to absorb heat and carbon, into the climate models. Cloud cover, too, has a strong moderating influence on the greenhouse effect, but it is difficult to characterize and incorporate into coarse-scale models. Even more difficult to model, and perhaps more important, are the living parts of the world—the forests, which store carbon and moisture, and the marine biota, which sequester carbon. Scientists look longingly to the day when enough is understood about these processes to include them in the models. Perhaps such a grand model can never be constructed, but the conceptual approach embedded in the attempt lays the cornerstone for earth system science.

OCEANS

The world's oceans are the atmosphere's partner in the physical climate system. Just as atmospheric chemistry fluctuates, so does ocean chemistry, though not in the same ways. While much is known about ocean circulation and its coupling to atmospheric currents and pressure, less is certain about its ability to store additional carbon or about how much heat it will store in response to rising surface temperatures.

The ocean is an immense reservoir of heat, holding the heat it absorbs from solar radiation longer than the land does. As the ocean water moves through its grand circulation scheme, heat is transferred vertically from the surface waters to the deep ocean and back, and horizontally from high latitude to low latitude and from longitude to longitude.

As heat is released by the ocean in a region remote from where it was absorbed, it interacts with the overlying atmosphere, moderating the daily and seasonal cycles and temperature on the earth's surface areas. Thus the ocean helps to shape the regional features of weather and climate. The episodic climate phenomenon known as the El Niño/Southern Oscillation, a change in atmospheric circulation that occurs irregularly every 2 to 7 years above the tropical Pacific Ocean, is the most notable example that a local disturbance in the balance between ocean and atmosphere can interact to cause an abrupt and dramatic change in the circulation of the tropical oceans and the global atmosphere. Kevin Trenberth, of the National Center for Atmospheric Research in Boulder, Colorado, and colleagues have shown that the hot and dry conditions in central North America in the summer of 1988 could have been triggered by unusual distributions in sea surface temperatures that occurred in the aftermath of the 1987 El Niño.

A critical unanswered question is, what is the ocean's role in storing the carbon dioxide added to the atmosphere by human activity? Preliminary calculations suggest that about half of the carbon dioxide added to the atmosphere by fossil fuel combustion and deforestation remains there. At least part of the carbon dioxide has been absorbed by the ocean, which holds 60 times as much carbon as there is in all of the atmospheric carbon dioxide.

The ocean's carbon largely resides at the bottom of the sea and has accumulated over billions of years. Photosynthetic plankton in the ocean's surface waters are consumed by other organisms; some of that carbon is returned to the atmosphere through respiration, and part goes into storage in the deep-sea sediment as detritus and shells or skeletons of marine organisms. The free-fall of organisms from the surface to the ocean floor and the subsequent release of carbon as deep ocean waters are slowly recycled up to the surface waters have a profound effect on the way carbon is apportioned throughout the earth system.

The movement of carbon through the earth system would be quite different if nothing lived in the ocean. If one could consider the influence of physics and chemistry alone, carbon dioxide in the surface waters would be evenly distributed. In fact, however, there is a distinct physical and chemical difference between the capacities of waters at different latitudes to sequester and release carbon dioxide. The food webs of organisms determine to what degree the carbon that is fixed photosynthetically will go back into the water and to what degree it will go into the deep ocean. In other words, the physics of the system that provides nutrients such as carbon, nitrogen, and other elements from the deep ocean to the surface and that moves surface waters from one location to another also influences the nature of the food web. In turn, the nature of the food web influences the partitioning of carbon dioxide.

James McCarthy, a biological oceanographer at Harvard University, believes that assumptions about the ocean's capacity for storing added carbon must be looked at carefully. What is it, he asks, that determines the capacity of the ocean today to absorb carbon? Why is it not half that amount, or twice that amount? How might the capacity of the ocean to absorb the carbon dioxide that is being released from fossil fuel combustion change in the future? What are the implications of this for the ocean carbon cycle? What would happen if the surface ocean conditions were to change?

Scientists have not yet answered these questions, but the record of the past provides some valuable clues in addressing them. The distinct correlation between the concentration of carbon dioxide and the surface temperature of the planet during the glacial cycles over the last 160,000 years must have involved the ocean. Researchers believe the carbon cannot move through any other reservoirs in the earth system efficiently enough over those time periods to account for these changes in carbon dioxide concentrations.

While there are many questions in urgent need of answers, in the last decade the ocean science community has developed new and powerful techniques for addressing them. Scientists have increased their understanding of the coupled nature of the atmosphere-ocean system, and of ocean physics and biogeochemistry. Increased computing and modeling capability improves researchers' ability to handle large data sets and to be able to put those data into forms that can be subjected to critical analysis.

The developments in remote sensing in the last decade have

been extraordinary. Until fairly recently, oceanographers based their studies of ocean processes on samples of ocean water gathered while aboard ships—an extremely slow, labor-intensive process. Ships move at roughly 10 knots, but weather patterns can move across the surface of the earth much faster. Indeed, much of the data collected from the ocean surface is biased because of problems of space and time scale. Now, satellites have made it possible to measure not only the ocean surface temperature but also how the surface currents are moving. Surface winds can be tracked with instruments aboard satellites, and the height of the ocean surface can be precisely gauged. These measurements reveal valuable information about ocean circulation. And, finally, the color of the ocean can be assessed to approximate the concentration of plankton pigment, and thus biological activity, at the ocean's surface.

LAND

Nothing seems more solid than a tract of land, and yet the plants and animals, the soil, and the life-supporting nutrients provided by that land make up a single interdependent unit—an ecosystem—that is dynamic on time scales ranging from days to seasons to years to millennia. Over days and seasons, the earth's plant communities absorb and release carbon in a breath-like rhythm. Over years and decades, ecosystems respond to the natural patterns of plant succession and occasional events such as El Niño or drought. At the far extreme, ecosystems on land change on time scales of tens to thousands of years according to the earth's glacial cycles.

Ecosystems function metabolically, producing and consuming many of the gases that drive the earth system. Plants capture energy from the sun and carbon dioxide from the atmosphere in their growth process. Terrestrial plants take up more than 100 billion metric tons of carbon each year and return approximately as much to the atmosphere as plants die and decay. This cyclical exchange involves 20 times the amount of carbon released through combustion of fossil fuels. Microorganisms in

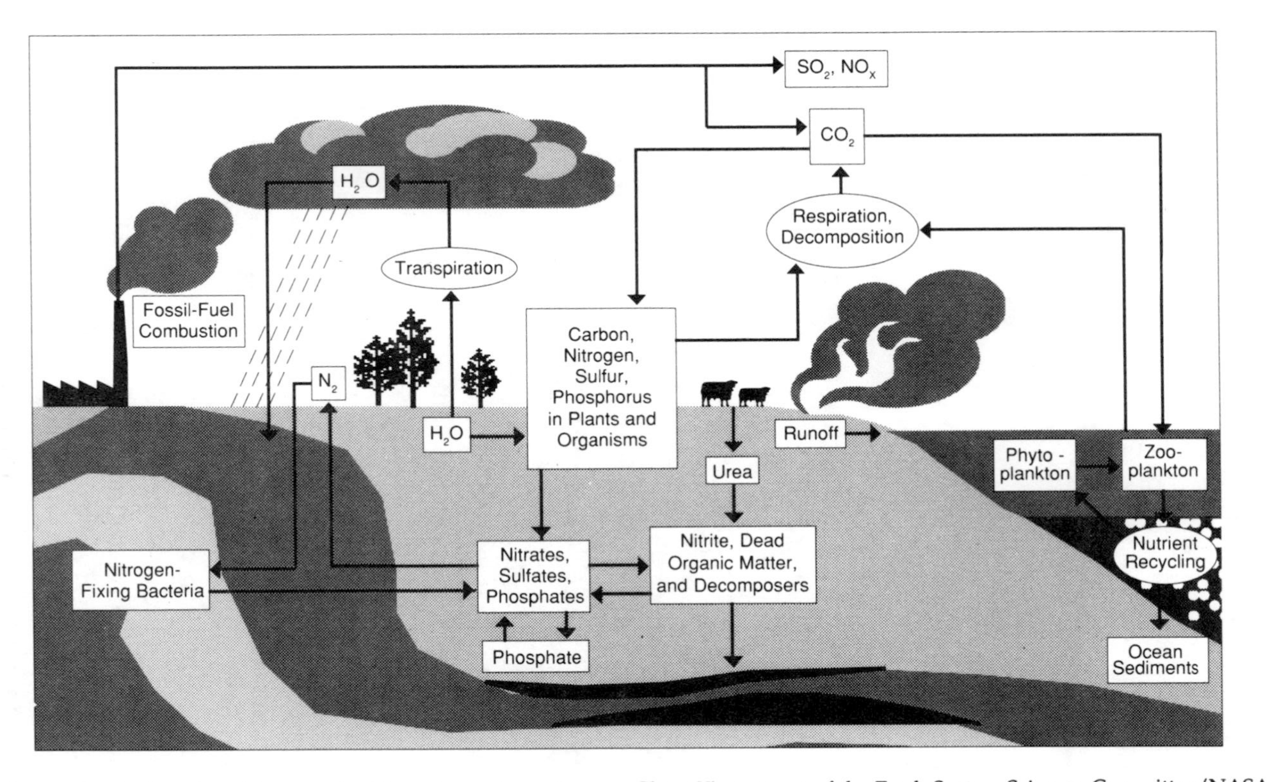

The biogeochemical cycles. Movement of key elements (carbon, nitrogen, sulfur, phosphorus, and others) through the earth system is critical to the maintenance of life. (From *Earth System Science: A Closer View*, report of the Earth System Sciences Committee/NASA Advisory Council, 1988.)

the soil release carbon dioxide and methane as end products and nitrogen-containing trace gases as by-products. As is the case with carbon, the amount of nitrous oxide cycled through terrestrial ecosystems is much greater than the amount released through combustion of fossil fuels.

Of the three main components of the earth system—atmosphere, oceans, and land—the land is the most heterogeneous. The earth's surface is a mosaic of different types of ecosystems ranging from arid desert to tropical forests to tundra to the more familiar temperate forests. Each harbors distinct plant and animal communities, and each uniquely contributes to the functioning of the earth system. Tropical rain forests, for example, with abundant moisture and high temperatures that facilitate exceedingly rapid plant growth and decomposition of dead plant material, cover about 7 percent of the earth's land area but contribute a much larger share of the world's annual turnover of biomass. At the other end of the spectrum, the cold temperatures in the tundra inhibit decomposition of plant material, and so the carbon in the biomass is stored there for long periods.

Though change is a quality intrinsic to all ecosystems, changes to the plant cover from agriculture, clearing of forests, and other human activities are not just another sort of change imposed on the background of natural variation. Rather, they profoundly alter the amount of light reflected back to the atmosphere from the land, the roughness of the land surface, which influences wind patterns, and the cycling of materials through the earth system.

For studies of the short-term dynamics of terrestrial ecosystems, biologists, like oceanographers and climatologists, have benefited from advances in satellite technology. One of the most important short-term dynamic effects is the seasonal variation in vegetation, which can be seen from space and recorded in snapshots. Some of these images show where plants are active at any given time and are extremely useful because the information can be accumulated daily, summed annually, and compared with measurements of the atmosphere. Peter Vitousek, a biologist at Stanford University, explains that results are particularly striking when the seasonal variation in the amount of light absorbed globally by vegetation is compared to the relative carbon dioxide concentration over the same seasons. "If you do this," Vitousek says, "not only do you see the biosphere inhale and exhale seasonally, but you actually see the distribution of the organisms over the surface of the earth engaged in that process."

In trying to assess what is in store for terrestrial ecosystems, researchers are drawn to the most recent instance when global climate changed on a massive scale: an ice age. The ice age is of particular interest in light of projections for the planet in the next century. Even during glaciation and the retreat of glaciers, which occurred much more slowly than the rate of warming projected for the planet in the next 100 years, the rate of change was so fast that only some species were able to adapt to the changes. Associations between species were severed. Eventually those species that survived recolonized into new communities, often in unfamiliar areas and in different combinations of members. As a result, many ecosystems were composed of wholly different combinations of species than are found anywhere today.

During the ice age, the major vegetation zones shifted thousands of kilometers from their current positions, and so the fraction of the earth's surface covered by specific types of vegetation also was altered substantially. What is in store for ecosystems in the future, and how these changes will feed back to other parts of the earth system, are open questions.

THE WATER CYCLE

It is easy to take water for granted. Rain, a lake, dew, waves crashing along a shoreline, snow, fog, a freshwater spring surrounded by desert palms—water in these and many other familiar forms means that life can be sustained. Nowhere else in the solar system does water currently exist in its liquid state; nowhere else has life taken root and flourished. Here water connects the various components of the biosphere, driving processes on land, sea, and air.

Like the other components of the earth system, water is

mobile on time scales ranging from the gradual advance of glaciers to the pelting of raindrops. It resides temporarily in oceans, groundwater, lakes, ice, and clouds and flows between them through rainfall and snow, evaporation from surfaces and through plants, and runoff across the earth's surface. Nearly every process in the earth system requires it. It sculpts the earth's topography, pushing vast amounts of debris ahead of advancing glaciers, compressing the land beneath mountains of ice. Soil particles caught up in river flows traverse great distances to the oceans and lakes, where they settle to the bottom and eventually harden into sedimentary rock. Water also destroys rocks, acting as a solvent in the weathering process or splitting them mechanically, pushing into crevasses where it freezes and expands.

Most aspects of the water cycle are poorly understood: There is simply too much of it in too many places for the many reservoirs, flows, and fluxes to be measured accurately. We do know that oceans hold the lion's share, over 97 percent, of the earth's water, followed by glaciers and ice caps. Lakes, rivers, and other surface water hold a mere one or two ten-thousandths of the global water stock.

People have affected the water cycle by constructing dams and reservoirs, which alter river flow and evaporation. Cities are built and paved, creating new patterns of runoff and preventing rainwater from entering the ground. Forests are cleared, reducing the ability of the soil and plants to retain water. People also consume water for drinking, cooking, and bathing and use it to irrigate their fields and to cool industrial plants. Such human actions raise the possibility that availability of water for future human use will be altered.

In light of the massive transformation under way in the global environment, water is of special interest because it exerts a strong moderating influence on the global climate system. Oceans, ice and snow, and clouds determine the earth's ability to reflect incoming radiation back to space, thereby helping to regulate temperature. In the form of water vapor—a greenhouse gas—water joins the other trace gases to absorb radiation leaving the earth's surface. Scientists are fairly certain that the water cycle, which transports and distributes most of the solar energy reaching the planet, will change in response to a warmer climate. Changes in factors such as the area of the earth covered by reflective polar ice or the abundance of clouds over the oceans, for example, would have a further effect on global temperature.

As temperature and hence evaporation from oceans and land increase, global precipitation is expected to increase by 5 to 10 percent. The timing and quantity of runoff may change, as will the amount of moisture stored in soil, with implications for world agriculture. Changes in vegetation in response to a warmer climate may profoundly affect patterns of evaporation and also whether precipitation seeps into the soil and groundwater for future use or runs off directly once it hits the ground. With current understanding, scientists cannot say how large the shifts in precipitation will be or where they will occur.

HUMAN INTERACTIONS

The recent furor over the changes humanity has wrought in the global environment since industrialization began invites the assumption that human alteration of the earth's landscape is a fairly recent phenomenon. In fact, many of our effects on the environment did not reach their global scale until the latter half of the twentieth century. But studies of many parts of the world suggest that as we extended our natural abilities with tools and later learned to cultivate plants, we became an effective agent of environmental change.

Ecologists Robert Peters, of the World Wildlife Fund, and Thomas Lovejoy, of the Smithsonian Institution, traced the record of human activity and its effect on terrestrial plant and animal life in several regions of the world. One of the areas they studied, the Mediterranean, provides a telling example.

Destruction of natural habitats around the Mediterranean began at least 7000 years B.C. Excavations show that by 6000 B.C., the bones of wild animals in kitchen refuse heaps were replaced by the bones of domestic sheep. In the fifth and fourth centuries B.C., forests began to dwindle as wood was harvested for fuel and construction. Around the Mediterranean, the researchers explain, humans have disrupted natural communities for so long that "it is difficult to determine which plants are natural or introduced, or what the original vegetation was like." Throughout the region, degradation of forested areas is so extreme that even if an area is protected, original vegetation often will not regenerate. Over the centuries, forests were converted to pasture, and grazing pasture was then replaced by thorny plants over enormous areas. Animal communities, displaced as their habitat disappeared, shrank in size and diversity.

Scientists have found patterns of human-induced change in other regions. Aborigines are thought to have walked into Australia from Indonesia about 40,000 years ago, when sea level was lowered during a glacial episode. Almost immediately, Australian vegetation became dominated by the fire-resistant eucalyptus tree. In Britain, habitat destruction over the last 3,000 to 4,000 years has caused 90 percent of its forest and most of its wilderness to vanish. In North America, as in Europe, marshes were drained, rivers dammed, and prairies plowed. And in Brazil's Atlantic forest, clearing began in earnest in the seventeenth century and continues today. From an original one million square kilometers, the Atlantic forest has been cleared until now, only fragments remain—less than 7 percent in any condition and less than 1 percent undisturbed.

The message in these examples is clear: With longer human occupation and greater population density, the influence of humans on other parts of the earth system grows. Now we know that human activities have become so pervasive that the effects are no longer local but are regional and even global in scale. Forest clearing is eliminating habitats where millions of species reside, acid rain is affecting lakes and streams in North America and Europe, and pollutants are changing the makeup of the atmosphere in ways that can affect climate and the protective ozone layer.

This awareness that humanity is an intrinsic part of the

earth system is causing a fundamental shift in the way science is pursued. No longer is it sufficient to explore only the physical dynamics of the earth system. This effort, daunting in itself, may be dwarfed by the effort to decipher the confounding behavior of *Homo sapiens,* the planet's most powerful inhabitant. Thus, as physical scientists join together to study, model, and predict changes on the earth's surface and in its atmosphere, their traditional focus on the physical and biological aspects of change is shifting to include the social sciences. For the first time, scientists from disciplines ranging from geochemistry to ecology are realizing that human action is the critical element in their studies. So potent is the human impact on the earth system that knowledge of physical processes ruling terrestrial or atmospheric change will be incomplete until scientists better understand the human dimensions of that change.

While studies in fields including economics, psychology, and communication provide an invaluable research foundation, they have, for the most part, focused on what determines and controls individual behavior. Roberta Balstad Miller, director of the Division of Social and Economic Science of the National Science Foundation, stresses that the study of human aspects of global change must consider not only individual behavior but also entire institutions—national laws and regulations, profit margins, transportation patterns, agricultural markets, and tax structures—that are significant for the environment. The research must also address the history of environmental change, dealing with human and institutional activities over long periods of time. "Research on the human dimensions of global change that ignores these factors would be nearly as inadequate as research that ignores the human dimension altogether," Miller said. "Will a social science research effort on global change be expensive? No question. But we must never forget that the costs of doing nothing are even greater."

The effort to discern the human causes of global change is complicated because the target changes over time: humans both act on and react to their environment. Using their unique

5

Humanity: An Agent of Global Environmental Change

The automobile, one of the most pervasive symbols of modern culture, serves as an apt metaphor for the ways in which humans change the global environment. Automobiles emit carbon dioxide that adds to the buildup of greenhouse gases in the atmosphere, nitrogen oxides that react in the atmosphere and rain down as acid deposition, and other gases and particulates that contribute to smog and local air pollution in our urban areas.

Today there are 500 million registered automobiles throughout the world, each of which burns an average of nearly 2 gallons of fuel a day. Automobiles consume one third of the world's production of oil. As the population grows, so will the number of automobiles. In addition, the average number of automobiles per person is going up, and the number of automobiles is increasing faster than the population, particularly in developing countries. Some estimate that if current trends continue, by 2025 there will be 4 times as many automobiles as there are today.

Paul Ehrlich, a population biologist at Stanford University, and John P. Holdren, an expert on energy and resources at Berkeley, pointed out in the early 1970s that the damage that humans cause to the environment is a product of three factors: the total number of people, how much each person consumes to maintain his or her standard of living, and how much environmental damage is incurred in producing the goods consumed. To pursue the automobile metaphor a moment longer, the same can be said for the environmental damage caused by automobiles or, for that matter, factories or other manifestations of society. For instance, the future environmental damage from automobiles depends not only on the total number of people owning them and the rate at which the number of automobiles per person increases but also on how much pollution each automobile emits. There are also signs that trends can be reversed. While advances in technology in the early 1900s made it possible for the automobile to preempt the horse as a mode of transportation, technology has also reduced the amount of pollutants emitted from each automobile. In the United States, for example, nitrogen oxide emissions from automobiles dropped substantially once catalytic convertors were required in all new cars.

Ehrlich and Holdren's equation embodies a simple and powerful truth. Our ability to change the global environment increases along with our numbers, our quest for and achievement of affluence, and our technological and institutional capabilities (which can also provide means to alleviate environmental damage). These intertwining factors, the product of decisions made by individuals and societies around the world, are the main forces driving change in the global environment.

POPULATION, AFFLUENCE, AND TECHNOLOGY: HOW HUMANS CHANGE THE ENVIRONMENT

A discussion of the human causes of global environmental change would not be complete without mention of the pressures that the sheer numbers of people inhabiting the earth place on the environment. Between 1950 and 1987, the global population doubled from 2.5 billion to 5 billion. This increase in the span

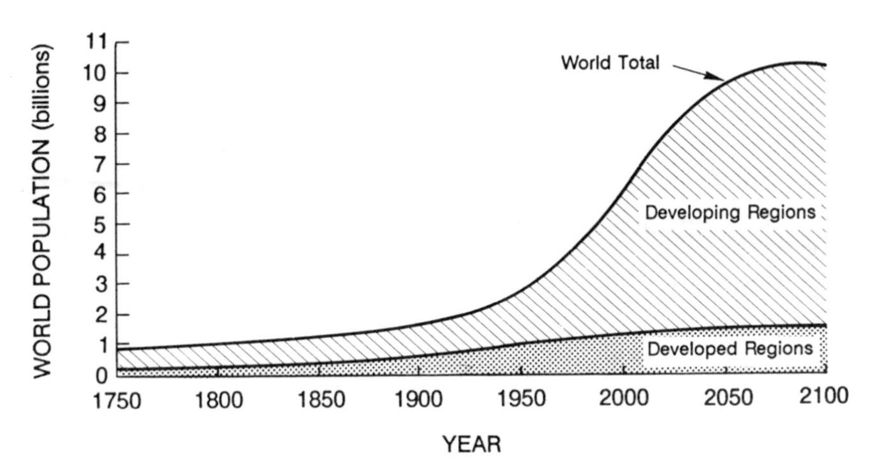

Population growth, 1750–2100. (Reprinted, by permission, from Thomas W. Merrick et al. 1986. "World Population in Transition," *Population Bulletin*, vol. 42, no. 2, Figure 1, p. 4. Copyright © 1986, Population Reference Bureau, Inc.)

of less than 40 years equaled the total increase in the world's population from the time the human species first emerged until the middle of the century. The United Nations estimates that by 2025 our numbers will increase to 8.5 billion before stabilizing at over 10 billion by the end of the coming century. Ninety-five percent of the population growth will be in developing countries.

This enormous increase in the numbers of people over the past century and the eventual stabilization in the total number of people are the product of the "demographic transition." With the industrialization that began in the last century in what is now the developed world, improvements in public health and medical advances caused death rates to decline. The improved socioeconomic conditions and access to health care and education that followed led to a drop in birth rates. When declining birth rates catch up with declining death rates, population eventually stabilizes. Indeed, the current net population growth in developed countries is generally very low.

In the developing world, in contrast, death rates have declined, but birth rates have not yet fallen enough to compensate.

The result is the very rapid population growth in developing countries that we see today. However, birth rates in many developing countries have begun to taper off in the past few decades, although the rate of decline differs among countries. In many African countries, for example, demographers expect that rates of population growth will not decline until at least the end of this century.

Demographers base their predictions of the world's future population on assumptions about how quickly birth rates will decline. If the rates do not decline as rapidly as assumed, the world's population will grow even faster than the United Nations currently predicts.

The explosion in the numbers of people inhabiting the earth is significant because, at the very least, each of these people requires food, clothing, and shelter. Growing numbers of people cut forests for land to grow food or graze cattle, with global consequences for the hydrologic cycle, the ability of the land to sustain agricultural productivity, and the earth's genetic resources as millions of plant and animal species are driven to extinction. Growing numbers of people in industrialized and industrializing societies use energy derived from fossil fuels to travel, produce goods, and apply advanced agricultural technologies, with global consequences for atmospheric concentrations of greenhouse gases, climate, and ecosystems.

But the impacts that the expanding numbers of people are having on the global environment are not solely a matter of the numbers of people; they also reflect how humans and the societies they live in use natural resources (which is, in turn, driven by economic and political structures as well as social values and norms). Societies use the air, water, and land as repositories for the waste products from industries that produce the goods that are consumed. They use fuel to drive cars, heat homes, and power industries. They convert raw materials such as wood and metals into consumer products and allocate land and fodder for animals that become food. If these resources are used in ways that emit excessive pollutants, clear forests, or deplete the soil of nutrients, the more societies consume to satisfy their standard of living, the greater the damage to the environment.

The earth's inhabitants do not exert equal pressure on the environment. Each person in an industrialized country uses far more natural resources and generates far more waste throughout his or her lifetime than a person in a developing country. The wealthy countries of the world constitute about one quarter of the global population but consume 80 percent of the world's commercial energy. Stated another way, each person in a developed country consumes on average about 32 barrels of crude oil per year, whereas a person in a developing country consumes only about 3.5 barrels. Energy use in many developing countries has been growing faster than population growth, but developing countries still consume 4 to 7 times less energy per person than their counterparts in the developed world.

Thus carbon dioxide, the predominant greenhouse gas produced as fossil fuel is burned for energy, is disproportionately emitted as a result of energy consumed in the industrialized countries. About 40 percent of all carbon dioxide building up in the atmosphere is emitted from seven wealthy countries in North America and Western Europe. These seven countries make up only 11 percent of the world's population.

While the numbers of people and their levels of affluence clearly influence the environment, the kind of technology used ultimately determines how much pollution is emitted or what kinds of and how much raw materials are used. In many industrialized countries, technologies to control emissions have in recent decades led to significant improvements in the air quality of many cities and the water quality in many lakes and rivers. Meanwhile, the amount of energy used to produce a unit of gross national product—known as the energy intensity—has declined in the industrialized world, by one fifth between 1973 and 1985. In the United States, according to John H. Gibbons, director of the U.S. Congress's Office of Technology Assessment, and colleagues, the gross national product grew 40 percent in these years while energy consumption remained constant.

Many who study technological trends perceive industrial

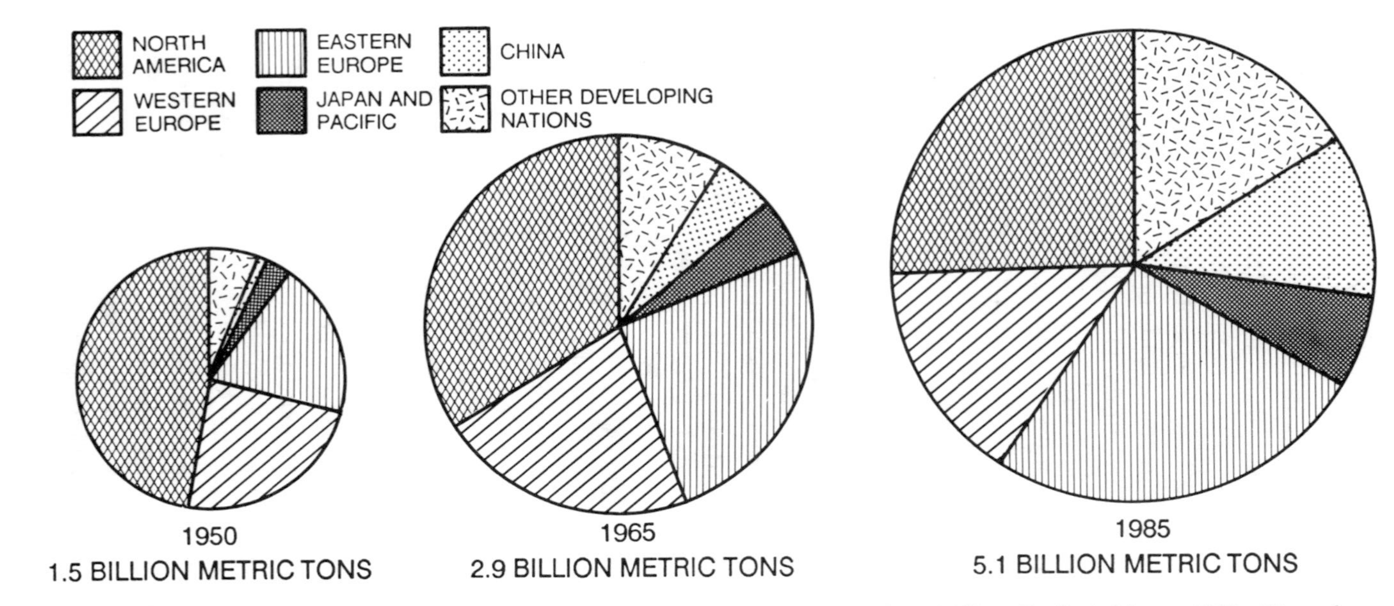

Developed nations are responsible for far more industrial emission of carbon dioxide—a major greenhouse gas—than are the developing nations. Total emissions have increased sharply since 1950. Data are from the World Resources Institute. (Reprinted, by permission, from William D. Ruckelshaus. 1989. "Toward a Sustainable World," *Scientific American*, Special Issue, "Managing Planet Earth," September, p. 170.)

economies to be undergoing a process of "dematerialization," a decline in the weight of materials and the amount of energy consumed in providing a given product or service. Undoubtedly, many consumer products such as automobiles, televisions, and calculators have become lighter and smaller. This trend means people can maintain a certain standard of living while using smaller amounts of raw materials and energy and creating less waste. But, as Robert Herman, of the University of Texas at Austin, and colleagues Siamak Ardekani, of the University of Texas at Arlington, and Jesse Ausubel, of The Rockefeller University in New York City, note, some trends may counterbalance these benefits. People migrating from urban to suburban areas require more roads, single-unit houses, and automobiles and demand more products as the definition of "basic necessities" changes. Such tendencies could offset some of the savings in waste, energy, and raw materials achieved through dematerialization.

Continued industrial development has serious implications for the future global environment, particularly if developing countries adopt older technologies that generate more pollution and consume more energy than more modern technologies. Clearly, though, economic growth is required to break the cycle of poverty and high birth rates that itself causes severe environmental damage to the soils, water, forests, and wildlife in many developing countries around the world. The experiences of developed and developing countries alike indicate that improved standards of living—adequate shelter, fuel, and clothing, as well as access to education, health care, and employment—led to a sustained reduction in birth rates. Whether this was because of improved access to birth control or greater rights for women or because with socioeconomic development parents no longer need many children to ensure financial security is a complex question. Nevertheless, the fact remains: As economies have grown and standards of living have improved, birth rates have declined.

CHANGING THE FACE OF THE EARTH: AGRICULTURE AND INDUSTRY

Humans do not tread softly on the earth. In fact, the effect of human activity is so pervasive that in just a few centuries the earth's surface has been transformed such that in many places its original state is difficult to imagine, much less restore. This transformation—wrought by the immensity of human numbers, by increasing affluence, and by technology—has captured the popular imagination in recent years as more people realize that their world, and their children's futures, may be jeopardized by unchecked environmental changes. The recognition that humans are a powerful agent in transforming the earth dates back much further, however. George Perkins Marsh in the nineteenth century, the Russian mineralogist W. I. Vernadsky in the first half of this century, and contributors to the now classic work, *Man's Role in Changing the Face of the Earth* (1956), edited by William L. Thomas, Jr., each expressed the ties between the global environment and human development.

B. L. Turner II, who heads the program on The Earth as Transformed by Human Action, at Clark University, points out that two types of transformations of the earth most affect the global environment: changes in how humans use the land, most notably for agriculture, and changes in industrial capabilities.

Over the past three centuries or so, as the area of land used for agriculture spread and the global economy expanded, humans were increasingly prone to adapt land to suit their own purposes. This control is most visible in the profound changes in the vegetation that covers the earth's surface. The land area devoted to cropland has expanded by 450 percent in the past 300 years, amounting to a world increase of about 12.4 million square kilometers. The net loss of forests due to human action from the time humans first began cutting trees amounts to 15 to 20 percent of the world's forested area, or about 8 million square kilometers. Also in the past three centuries, the population of the 40 largest urban areas increased about 25-fold, with an even greater increase in estimated urban area. Such changes have drastically altered the earth's surface, as well as the energy balance, the hydrologic cycle, the emissions of biologically produced gases to the atmosphere, and the vegetation.

Industrial development, too, has a potent effect on a global scale, through pollutants and other materials that are introduced into the biosphere, atmosphere, and oceans. Industrial development has vastly improved human well-being, but there have been unexpected side effects. Chlorofluorocarbons (CFCs) were invented around 1930 and were hailed, for instance, as safe alternatives to ammonia and other coolants that had a tendency to leak and cause explosions when used in home refrigerators. No one could foresee the repercussions of these seemingly safe, stable, and inexpensive compounds on the global environment. Certainly, no chemist imagined that CFCs destroy the ozone layer about 25 miles above the earth, which shields us from solar ultraviolet radiation, or that CFCs contribute roughly 25 percent of the warming caused as greenhouse gases build up in the atmosphere.

Other examples abound of what Paul Gray, president of the Massachusetts Institute of Technology, calls the "paradox of technological development." Writing in the 1989 National Academy of Engineering publication *Technology and Environment*, Gray observed that "New technology will be applied in ways that transcend the intentions and the purposes of its creators, and new technology will reveal consequences that were not anticipated." A few decades ago, asbestos fibers were hailed as a breakthrough for fireproofing ships, schools, and office buildings, but today they are known to cause lung cancer. The "green revolution" that improved food production around the world and enabled countries such as India to become net exporters rather than importers of food has contributed to the chemical pollution of rivers, lakes, and seas because it relies heavily on the use of insecticides and fertilizers.

Energy is fundamental to industrial development, and the sources used to derive that energy have had particular implications for the environment. With the Industrial Revolution in the early 1800s, fuel wood and water power gave way to coal

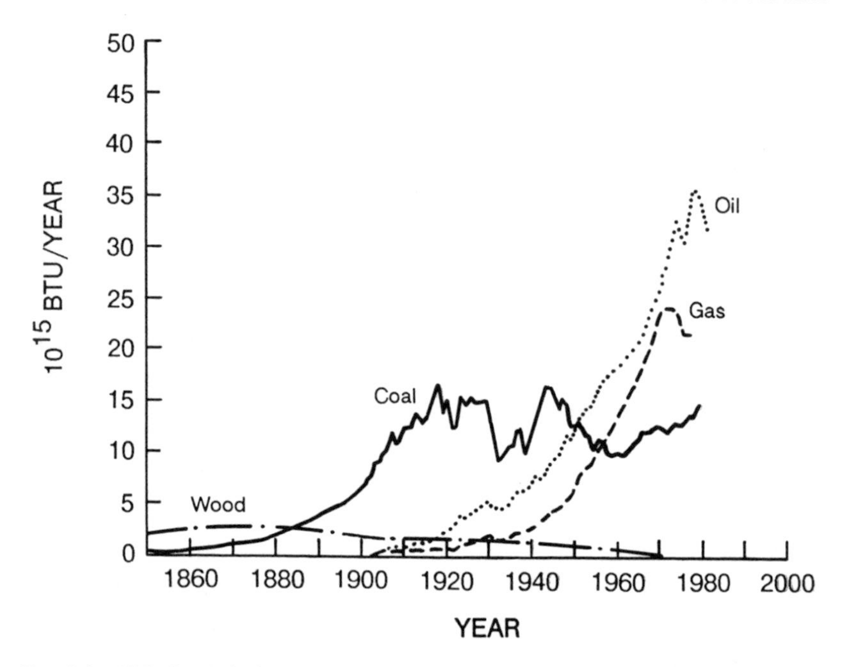

Trend for U.S. fossil fuel consumption (in British thermal units, BTUs) since 1850. Data for 1850 to 1880 from U.S. Bureau of the Census (1975); data from 1880 to 1932 from U.S. Geological Survey, Yearbooks (1880-1932); data from 1933 to 1980 from U.S. Bureau of Mines, Mineral Yearbooks (1933-1980). (Reprinted from National Research Council. 1986. *Acid Deposition: Long-Term Trends*. National Academy Press, Washington, D.C.)

and coal-based technologies such as iron smelting and steam engines. By the early 1900s in the United States, coal, which generates large amounts of carbon dioxide and other pollutants, supplied most of the energy for the nation's factories. Industrialized countries currently generate between 20 and 30 percent of their energy from coal. Natural gas, which currently provides about one fifth of the world's commercial energy, generates fewer pollutants and less carbon dioxide than any other fossil fuel.

Today, fossil fuels (oil, natural gas, and coal) supply 88 percent of the world's commercial energy, and nuclear energy provides the rest. In many poor countries, noncommercial fuels such as wood, dung, and crop waste still provide much of the energy for cooking and heating. Clearly, the future of the global environment depends to a large extent on the sources of energy for current and future societies and the amount of energy consumed.

THE FUTURE

The global consequences of industrial development as practiced today cannot be denied. Waterways, soils, and forests bear the signature of acid deposition; the annual hole in the ozone layer marks human-induced depletion of this protective gas. These environmental changes may be unacceptable costs of progress, and yet development must continue, albeit in a less damaging form. The gap between rich and poor can be narrowed only by the economic development necessary to meet the needs and aspirations of the world's poor today, as well as the needs and aspirations of tomorrow's growing population. The challenge, in the words of Jim MacNeill, director of the Environment and Sustainable Development Program at the Institute for Research on Public Policy in Ottawa, Canada, and a contributor to *Our Common Future*, the 1987 report of the U.N. World Commission on Environment and Development, is to "emphasize the possibilities for a 'new era of growth'—not the type of growth that dominates today but sustainable growth, growth based on forms and processes of development that do not undermine the integrity of the environment on which they depend."

In one such approach, Robert Frosch and Nicholas Gallopoulos, of General Motors, suggested in the 1989 Special Issue of *Scientific American* that the traditional model of industrial activity—in which individual manufacturing processes take in raw materials and generate products to be sold and waste to be disposed of—should be transformed "into a more integrated model: an industrial ecosystem." Like its analog in the biological ecosystem, such an industrial ecosystem would optimize consumption of energy and materials, minimize waste generation, and use the effluents of one process—whether fly and bottom ash from electric-power generation or discarded plastic containers from consumer products—as raw materials for another process.

Just as humans can transform the face of the earth, we can also mitigate some of the adverse effects. Methods exist that could shape the future environment, including regulatory control and market adjustments in which prices of goods account for the environmental damage they produce (the price of gasoline could, for instance, reflect the costs associated with smog, global warming, or acid deposition), as well as changes in social values and norms. William Ruckelshaus, former head of EPA, notes that history offers examples in which nations have mobilized to synthesize short-term interests with longer-term goals. During World War II, for instance, when people believed that the nation's survival was at stake, seemingly impossible political and economic changes were accomplished in a remarkably short time. The entire U.S. population was mobilized: Work patterns changed, prices and supplies of goods were manipulated and controlled, and the nation's industrial plants were swiftly reorganized.

Even though government agencies and international organizations are structured to deal with environmental and economic issues separately, these issues are in fact interlocked. "Global warming is a form of feedback from the earth's ecological system to the world's economic system," MacNeill observes. "So are the ozone hole, acid rain in Europe and eastern North America, soil degradation in the prairies, deforestation and species loss in the Amazon, and many other environmental phenomena." These phenomena are the faces of global change.

THE FACES OF GLOBAL ENVIRONMENTAL CHANGE

6

Global Warming

As the weeks of 1988's summer drought and stifling heat dragged by, a momentous shift took place in the public's attitude toward the global environment. Suddenly, it seemed, everyone knew and cared about a scientific principle long of deep concern to scientists studying the earth system. This principle, known as the greenhouse effect, explains why gases produced by human activity will probably cause the earth's average temperature to increase within the lifetimes of most people living today.

The 1980s were the warmest decade recorded on a global basis, but scientists are still uncertain, and will be for years, whether the warm spell was a normal climatic fluctuation or a response to the billions of tons of carbon injected into the atmosphere each year by human activities. Scientists working in climatology and related fields say that the insulating effects of the greenhouse gases should be clear to all of us within a few decades, and possibly by the end of the 1990s.

One cannot infer from a specific summer that global warming has begun, though a warmer climate would change the probabilities for heat waves and possibly for strong hurricanes. The weather events of 1988 did, however, convey an idea of the kinds of environmental and commercial effects we could expect if the current predictions about global warming come to pass. The North American corn crop was stunted by drought in the grain belt, and productivity fell below consumption (probably for the first time in U.S. history), so that no grain was added to the nation's reserves. Water levels in thoroughfares like the Mississippi River dropped so low that barges and their cargoes were stranded for weeks. Forest fires burned uncontrollably in America's great natural parks, a superhurricane threatened America's Gulf Coast, and floods in Bangladesh led to the deaths of 2000 people and drove millions of others from their homes. These and other extreme weather events over the course of a single summer highlighted for billions of people that human society is highly vulnerable to extremes in the weather.

GREENHOUSE GASES

Although there may be questions about the causes of a specific drought or flood, there is no controversy about some basic facts about our atmosphere. Trace gases such as water vapor, carbon dioxide, methane, chlorofluorocarbons, tropospheric ozone, and nitrous oxide create a greenhouse effect by trapping heat near the earth's surface, and the concentrations of many of these gases are increasing in the atmosphere. Because of these increases, the gases are expected to trap more energy at the earth's surface and in the lower atmosphere, in turn causing increases in temperature, changes in precipitation patterns, and other as yet unpredictable changes in the global climate.

The principle of the greenhouse effect explains the cold climate of Mars (where water vapor, a highly efficient greenhouse gas, is virtually absent), the hot climate of Venus (where the atmosphere is thick with carbon dioxide and conditions are so hot that life as we know it could not survive), and the moderate climate here on earth. Scientists have known for decades that a buildup of carbon dioxide in the atmosphere could warm the earth's climate. They have also known that atmospheric concentrations of carbon dioxide alone have increased by about 25 percent since coal, oil, and gas became the primary sources of energy to fuel the Industrial Revolution. Carbon dioxide concentrations are currently increasing by about 0.4 percent each year.

After water vapor, carbon dioxide is the most plentiful and effective greenhouse gas. It occurs naturally but is also produced in great quantity during the combustion of fossil fuels, particularly coal. When the fuel is burned, its carbon is oxidized to carbon dioxide and released. Carbon dioxide also is released as forests are cleared and the organic matter is burned or allowed to decay. These human activities are injecting almost 6 billion tons of carbon into the atmosphere each year. By comparing this figure with the actual increases in concentrations of carbon dioxide (about 3 billion tons annually), scientists presume that about half of the carbon injected into the atmosphere is being absorbed by oceans and plant life and about half remains in the atmosphere.

Only in the last decade have scientists become aware that other, trace greenhouse gases can also be important contributors to global warming. Concentrations of many of these trace gases are known to vary naturally, but there is widespread agreement that human activities are contributing to the current increases.

Molecule for molecule, the following trace gases absorb infrared radiation much more effectively than carbon dioxide does. Because their concentrations in the atmosphere are much lower than that of carbon dioxide, their individual effect is much smaller. Their combined effect, however, is likely to be equal to or greater than that of carbon dioxide alone.

Methane (CH_4). Methane, also known as natural gas, is produced through bacterial activity in bogs and rice paddies, and in the digestive tracts of ruminative animals and insects such as termites. Most atmospheric methane comes from biological sources. It is present today at roughly 1.7 parts per million and is increasing at a rate of about 1.1 percent each year. Analysis of gas bubbles trapped in glacial ice shows that the rise in methane levels parallels the growth of the human population. Per molecule it is 25 times as effective as carbon dioxide at trapping heat.

Chlorofluorocarbons (CFCs). Chlorofluorocarbons are a group of synthetic compounds used in refrigeration, insulation, foams, and other industrial purposes. Apart from their role as greenhouse gases, when CFCs rise to the upper atmosphere, or stratosphere, they release free chlorine, which then catalyzes the breakdown of ozone, the protective layer that shields the earth from ultraviolet radiation. The two most prevalent CFCs are CFC-12, which per molecule has 20,000 times the capacity of carbon dioxide to trap heat, and CFC-11, which has 17,500 times the capacity. Both of these compounds are long-lived and are increasing in the atmosphere at a rate of about 5 percent per year. The Montreal Protocol, an international agreement adopted in 1987 to limit the production of CFCs, will slow but not eliminate the rate of increase.

Nitrous oxide (N_2O). Nitrous oxide is produced naturally, through microbial action in the soil, and in response to the spread of agriculture, the burning of timber, the decay of crop residues, and the combustion of fossil fuels. Agricultural use of mineral fertilizers containing nitrogen presumably accelerates its rate of release. Atmospheric concentrations of nitrous oxide are increasing by about 0.25 percent per year. It has a long residence time in the atmosphere, and so concentrations would increase for more than 200 years even if emission rates were to freeze at current levels. Scientists believe nitrous oxide levels in the year 2030 will be about 34 percent more than preindustrial levels. Per molecule, this trace gas has 250 times the capacity of carbon dioxide to trap heat.

Tropospheric ozone (O_3). In the stratosphere, ozone shields the planet from ultraviolet radiation; nearer the ground in the troposphere, the moisture-rich atmospheric layer below the stratosphere, it is an effective greenhouse gas. It is produced through reactions involving hydrocarbons and nitrogen oxides, all released through the combustion of fossil fuels used by motor vehicles and in industry. Concentrations of tropospheric ozone appear to be increasing at many locations in the Northern Hemisphere. Results from studies of the Amazon River basin indicate that tropical forests act as a sink for ozone; if so, their continued destruction could have a significant effect on regional ozone balances.

Although scientists have considerable confidence in the principle of the greenhouse effect and the measurements of increasing greenhouse gases in the atmosphere, two key questions remain surrounded by uncertainties: How quickly will the climate change, and by how much?

THE CLIMATE'S RESPONSE TO GREENHOUSE GASES

Using three-dimensional mathematical models of the climate system, scientists draw a number of inferences about what conditions might be like in the future. The most likely conditions include significant cooling of the stratosphere, warmer surface temperature (which would be felt disproportionately at high latitudes), and changes such as rising sea level, reductions in sea ice, and increases in total global precipitation (which again would be nonuniformly distributed around the globe). They also speculate that summers in the mid-continents would be much dryer than they are today.

These responses to increased greenhouse gas concentration, as well as their scientific uncertainties, were described by Jerry Mahlman, director of the Geophysical Fluid Dynamics Laboratory of the National Oceanic and Atmospheric Administration (NOAA) in Princeton, New Jersey, at the 1989 Forum on Global Change and Our Common Future. Mahlman outlined a list of responses, an earlier version of which appeared in the

1987 National Research Council report, *Current Issues in Atmospheric Change* (see box). The estimates shown reflect assumptions about future concentrations of trace gases based on current trends.

The results of literally millions of measurements and calculations over the past century indicate that the earth is likely to experience a significant climate change during the next few decades. The models predict that because of carbon dioxide and other gases that have built up in the atmosphere since 1860, the earth is probably already committed to a 0.5° to 1.5°C increase in average global temperature. If current emissions trends continue, the combined greenhouse effect of all trace gases would commit us to an "effective carbon dioxide doubling"—the point where carbon dioxide and other trace greenhouse gases combined trap the same amount of energy that carbon dioxide would trap alone if its concentration doubled from the preindustrial level—possibly as early as 2030.

Although the climate models are intricate and require massive amounts of computer time, they are stark, simplistic representations of the complex realities of the real climate system. It is difficult, for instance, to include cloud cover in the models, even though clouds may amplify or moderate the greenhouse effect. Most of the models do not adequately include the dynamics of ocean circulation, an essential determinant of carbon dioxide concentrations in the atmosphere. Nor can the models incorporate the entire range of uncertainties about potential responses of the earth system—the possibility, for instance, that an increase in temperature could alter cloud cover or increase the rate at which soil bacteria break down dead organic matter and consequently accelerate the biological contribution of carbon dioxide to the atmosphere, or the possibility that climate change could trigger a dramatic shift in ocean circulation that would completely alter temperature and precipitation patterns.

In spite of these uncertainties imposed by both the practical computational limits of the models and the incomplete understanding of the earth system, scientists cautiously predict how much global average temperatures would rise with an effective

General circulation models predict several important climate changes due to increased greenhouse gases. Jerry Mahlman interprets confidence levels for current models according to the following guidelines: "Virtually certain" means there is nearly unanimous agreement within the scientific community that a given climatic effect will occur. "Very probable" means greater than about a 90 percent (9 out of 10) chance, and "probable" implies more than about a 67 percent (2 out of 3) chance. "Uncertain" refers to an effect that has been hypothesized but for which there is a lack of appropriate modeling or observational evidence.

- *Large Stratospheric Cooling* (virtually certain)

A reduction in upper stratospheric ozone by chlorine compounds will mean less absorption of solar radiation and thus less heating. In addition, increased concentrations of trace gases will increase the loss of infrared radiative heat from the stratosphere. The combination of decreased heating and increased cooling will markedly lower upper stratospheric temperatures, perhaps by 10° to 20°C.

- *Global Mean Surface Warming* (very probable)

For a doubling of atmospheric carbon dioxide (or its radiative equivalent from all of the greenhouse gases), the long-term global mean surface warming is expected to be in the range of 1.5° to 5.0°C. The most significant source of uncertainty arises from the difficulties in modeling the feedback effects of clouds. The actual rate of warming over the next century will be governed by the rate at which concentrations of greenhouse gases increase, natural fluctuations in the climate system, and the intricate responses of the slowly responding parts of the climate system, such as oceans and glacial ice.

- *Global Mean Precipitation Increase* (very probable)

Increased heating of the earth's surface will lead to increased evaporation and therefore to greater global mean precipitation. Nonetheless, precipitation might well decrease in many individual regions.

- *Northern Polar Winter Surface Warming* (very probable)

As the sea ice boundary shifts toward the poles, the models indicate that winter surface temperatures in polar regions would be much warmer than they are now. Scientists calculate that the greater fraction of open water and thinner sea ice in winter will cause the northern polar surface air to be more than 10°C warmer than the current climate, or 3 times the global mean warming.

- *Reduction of Sea Ice* (very probable)

As the climate warms, there will be less total sea ice because of warming at high latitudes of the Northern Hemisphere. However, the NOAA Geophysical Fluid Dynamics Laboratory reports preliminary findings that new model results incorporating a fully interactive ocean indicate little warming at high Southern Hemisphere latitudes over the next century. This would mean little change in sea ice cover there.

- *Northern High-Latitude Precipitation Increase* (probable)

As the climate warms, the increased poleward penetration of warm, moist air may increase the annual average precipitation and river runoff in high latitudes.

- *Summer Continental Dryness/Warming* (probable)

Several model studies have indicated that soil moisture in the mid-latitude continental interiors may decrease during summer. This drying is caused mainly by an earlier termination of snowmelt and rainy periods and thus an earlier onset of the normal spring-to-summer reduction of soil moisture.

- *Rise in Global Mean Sea Level* (probable)

It is likely that sea level will rise as sea water expands in response to the warmer future climate. Far less certain is how much sea level will be affected by possible melting of glaciers and calving of land ice.

• *Regional Vegetation Changes* (uncertain)
Climatic changes in temperature and precipitation of the kinds indicated above must inevitably lead to long-term changes in surface vegetation. The exact nature of such changes and how they might affect the climate in turn remain uncertain.

• *Tropical Storm Increases* (uncertain)
A number of scientists have suggested that a warmer, wetter atmosphere could lead to more frequent and more intense tropical storms, such as hurricanes. However, tropical storms also are governed by other factors such as local wind structure. So far, this effect has not been satisfactorily addressed in the coarse-resolution climate models because tropical disturbances are relatively small.

carbon dioxide doubling. Many assessments indicate a range of estimates between 1° and 5°C.

To appreciate what the projected warming really means, consider the numbers involved. When scientists say that on average the global temperature could increase by a few degrees centigrade, they are talking about a very large increase and a tremendous amount of heat. The current average global temperature is about 14°C (57°F). A 3°C rise would create conditions that some organisms have not had to contend with in the last 100,000 years. If the temperature rises 4°C, the earth would be warmer than at any time since the Eocene period, 40 million years ago. In the midst of the last glaciation, when much of North America was covered by ice, the average temperature of the earth was only about 5°C colder than it is now. Thus, what seems to be a very small average temperature change can have a very dramatic effect. Moreover, the projected rate of warming is 15 to 40 times faster than the natural warmings after the major ice ages and much faster than what most species living on the earth today have had to face.

The climate model results shown in the box are based mainly

on hypothetical and mostly instantaneous and large changes in concentrations of greenhouse gases. In fact, concentrations of the gases are increasing gradually. Initially, much of the excess heat is absorbed into the oceans, but understanding of the complex interactions between the atmosphere and ocean is incomplete. We can expect that natural, decadal-scale climatic fluctuations due to interactions between the atmosphere and oceans will continue to occur. Mahlman points out that the midwestern drought in the 1930s and the high water levels of the Great Lakes in the 1980s are good examples of the results of such fluctuations. Until such fluctuations can be understood and predicted, it will be difficult to discern the specific signals of more long-lasting climate change as they evolve. Detecting the signals of climate change becomes even more difficult when smaller regions and/or shorter periods of time are considered.

The enormous consequences of the various effects of global warming and the rising clamor for clarification continue to spur the scientific community to refine their mathematical models. Despite scientific uncertainties, these computer models are the only tools available to researchers as they struggle to estimate to what extent economic and social actions to reduce future emissions of greenhouse gases can limit the predicted changes in climate. Stephen H. Schneider, a climatologist at the National Center for Atmospheric Research in Boulder, Colorado, and Norman Rosenberg, director of the Climate Resources Program for Resources for the Future, note that another decade or so of observations will enable scientists to assess how well present estimates predicted the sensitivity of climate to increasing trace gases. But, they add, "While scientists debate, the real climate system continues to perform the experiment for us."

All of the predictions about climate change are based on only five models (although there are many attempts to model portions of the earth system on more limited scales of time and space). The five models are the NASA/Goddard Institute for Space Studies (GISS) model, the National Center for Atmospheric Research (NCAR) model, the NOAA Geophysical Fluid

Dynamics Laboratory (GFDL) model, the model developed at the Oregon State University (OSU), and the model developed by the United Kingdom Meteorological Office (UKMO).

These are general circulation models (GCMs) that predict the ways in which temperature, humidity, wind speed and direction, soil moisture, sea ice, and other climate variables evolve through three dimensions and over time. They use mathematical equations to express the basic physical, chemical, and biological processes that govern the global climate system.

The general circulation models agree that change is in the works and that weather systems worldwide are sensitive to increases in greenhouse gases. Their calculations reveal that disruption is all but inevitable and that a wide range of consequences is possible. There are differences between them, however. Some versions of the GISS and GFDL models now include scenarios of gradual addition of greenhouse gases into the atmosphere, whereas the others assume a massive, one-time doubling of the gases. The GFDL and OSU models attempt to include ocean processes. The GFDL model indicates that some remarkable effects can occur when an active circulation is included. For example, the presence of upwelling circulation in the circum-Atlantic Ocean acts to delay surface warming there for extended periods, perhaps centuries.

The models do not necessarily agree on specifics. All project that average precipitation over the globe will increase significantly but differ on what the regional effects would be. The climate system is so complex and so vast that it is a mind-boggling proposition to decipher the interactions and balances among its myriad components. So far, even the most sophisticated coupled atmosphere-ocean model omits important features such as the biological interactions. Also, current computer power is insufficient to resolve many climatically significant phenomena. Most modelers believe reliable predictions from this crucial tool are 10 or 20 years away and that until (and even after) it exists surprises are likely.

No computer can handle all of the calculations required

to simulate the complexity of the atmosphere. To compensate, scientists make calculations for areas encompassed by widely spaced points that form a three-dimensional grid at and above the earth's surface. In the current models, spacing, or resolution, of the grid is 300 miles, or 5° latitude. This kind of spatial resolution means that for purposes of regional analysis, Panama, for instance, does not exist, and neither does Japan. Nor does it accurately reflect the influence of factors like clouds, because they occur over a much smaller area.

If resolution were increased to 2.5° latitude, the cost of running the computer models would be more than 10 times greater. At a resolution of 1° latitude by 1° longitude, modelers could calculate effects over an area 60 miles on a side—a useful size for studying regional effects on natural ecosystems, agriculture, and water supplies. This would require 500 times as much computer time, at great expense. Thus the demands of policymakers will outstrip the ability of climatologists to deliver answers for probably the next two decades.

How well can the models simulate climate? As Schneider explains, "Perhaps the most perplexing question about climate models is whether they can ever be trusted enough to provide grounds for altering social policies, such as those governing carbon dioxide emissions." How can models so fraught with uncertainties be verified? Schneider explains that there are three main tests that together can provide evidence about a model's credibility: whether the model can simulate today's climate, especially the large temperature swings of the seasonal cycle; whether the model can realistically simulate an individual physical component of the climate system, such as cloudiness; and whether the model can simulate long-term climate changes by reproducing the varied climates of the ancient earth, or of other planets. How the models perform against such known standards is constantly being reappraised by their users.

The success of the models in passing these tests and the ability of different models to have similar results show that the models are getting better at predicting climate change, though there is much room for improvement in coming decades.

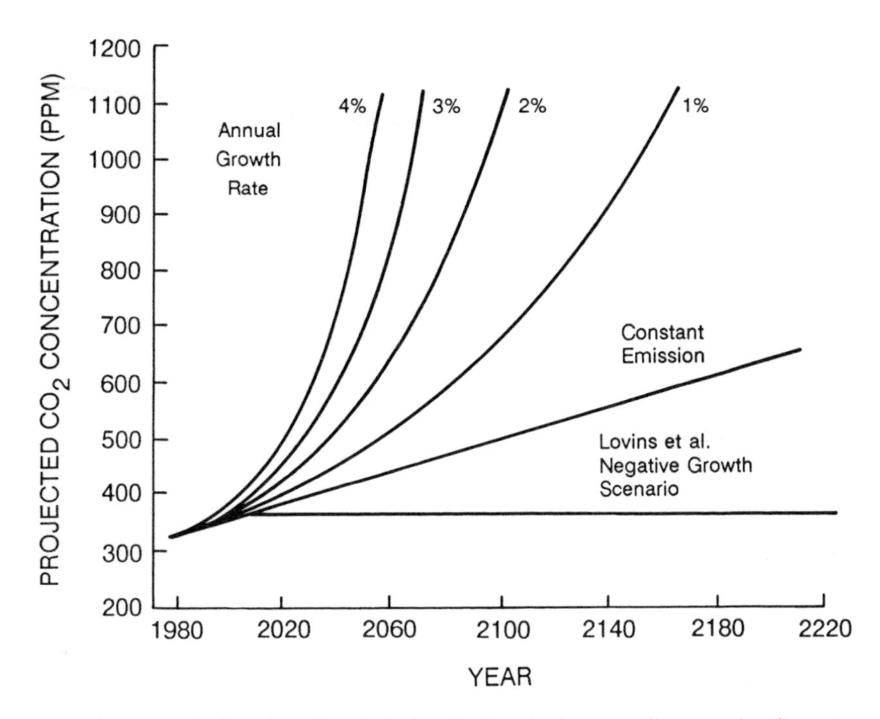

The extent to which carbon-dioxide-induced climatic change will prove significant in the future depends, of course, on the rate of injection of carbon dioxide into the atmosphere. This depends, in turn, on behavioral assumptions as to how much fossil fuel burning will take place. (This graph neglects biospheric effects such as carbon dioxide emissions due to deforestation.) Since the end of World War II, a world energy growth rate of about 5.3 percent per year occurred until the mid-1970s, the time of the OPEC price hikes. Rates have come down substantially since then and hardly grew at all in the early 1980s. The figure shows projected carbon dioxide concentrations for different annual growth rates in fossil energy use, including one for the assumption that no increase in fossil energy use occurs (constant 1975 emission) and even a "negative growth scenario" (Amory B. Lovins et al.) in which energy growth after 1985 is assumed to be reduced by a fixed amount (0.2 terawatts [TW] per year, which is about 2 percent of present demand) each year. (Reprinted, by permission, from Amory B. Lovins et al. 1989. *Least-Cost Energy: Solving the CO_2 Problem*, Figure 1.1, p. 10. Copyright © 1989, Rocky Mountain Institute. As adapted from Stephen H. Schneider. 1989. *Global Warming: Are We Entering the Greenhouse Century?*, Figure 6, p. 100.)

DO WE KNOW ENOUGH TO ACT?

The rising concentrations of greenhouse gases in the atmosphere are a direct response to our actions as we conduct our lives, drive our vehicles, grow our food, and run our industries. We are transforming the environment that sustains us.

How much warmer the climate becomes, and how quickly the warming occurs, depend on whether societies decide to act to slow emissions of carbon dioxide and other trace gases. Scientists can provide raw material that can be analyzed before such decisions are made, but whether to act is a social judgment, not a scientific one.

The question is then, do we, and those who set the world's environmental, economic, and social policies, know enough to decide whether to slow the rate of greenhouse gas emissions and, thereby, the rate of global warming? While acknowledging the many uncertainties, many members of the scientific community believe the answer is a guarded "yes," particularly because the more rapid the change in climate, the more difficult it will be for societies and ecosystems to adapt.

Many effects of global warming, such as those on agriculture, will be felt unequally around the globe. Researchers can predict with a fair degree of confidence that changes in temperature, precipitation, and soil wetness will affect agriculture, improving the competitive advantages of some crops and regions and lessening others, but they cannot say with certainty which ones. They can pinpoint which coastal areas would be most affected by a rise in sea level as glaciers melt and the oceans expand in response to the extra heat. But the faster change occurs, the greater the likelihood of unforeseen consequences. As Schneider notes, "Quite simply, the 'bottom line' of the evolving greenhouse gas build-up is that we insult the environment at a faster rate than we can predict the consequences, and that under these conditions, surprises are virtually certain."

The following chapters describe the sweeping changes under way or predicted in the global environment, changes caused by humans as we attempt to satisfy the needs of the world's growing population. Some of the direct consequences of global warming for society—the effects on food supply and the impacts on coastal areas—are discussed in the following two chapters. Other issues discussed in subsequent chapters—global environmental issues in their own right—are also intricately tied to global warming: Acid deposition is caused by fossil fuel combustion, as is the major share of the increase in greenhouse gases; destruction of the ozone layer that shields us from the sun's harmful radiation is caused by industrially produced chlorofluorocarbons, also a powerful greenhouse gas; and the large-scale felling of the world's tropical forests contributes to the increase in atmospheric carbon dioxide, in addition to eradicating the habitat for millions of plant and animal species. Many researchers fear that global warming will accelerate the pace of species extinction as plant and animal communities are torn apart by the stresses of adapting to a quickly changing climate.

Each of these problems demands at least attention and possibly action even if the projected global warming never occurs. Schneider is a vocal proponent of what he has dubbed the "tie-in" strategy, in which individuals, firms, and nations would take steps to slow down the rate of buildup of greenhouse gases and at the same time tackle other societal problems. As insurance against the surprises that would be more likely the faster the climate change occurs, he urges accelerated testing of alternative non-fossil fuels, development of strains adapted to wider climate ranges, adding flexibility to the management of water systems, and coastal planning to deal with rising sea level and storm surges. Just one initiative—energy conservation—could reduce the impact of many immediate problems. More efficient fuel use would cause air pollution to decline, cut acid rain, lessen the dependence of many nations on unreliable sources of oil (thereby increasing security), and improve the competitiveness of manufactured goods as the cost of producing them drops along with energy use.

Failure to take steps may force us and other living things to adapt to a much larger dose of change than if we act today to slow down the change or to invest to make future adaptations easier. Says Schneider, "Choosing to wait until the greenhouse effect signal has clearly been detected in the climatic record is not a cost-free delay. It is a basic gamble with our environmental future."

7

Food, Water, and Changing Climate

As scientists work to improve their predictions of change in the earth's climate, a long-standing question assumes ever-greater currency: How many people can the planet support without using up our natural resources and forever undermining the earth's ability to support people in the future? In other words, what is the carrying capacity of the earth?

Today, assuming equitable distribution to the 5 billion people all over the world, the earth certainly provides enough food for an adequate diet. This fact, however, conceals a distressing paradox: In recent years, although the earth produced record amounts of grains, half a billion people were seriously malnourished. One reason is the unrelenting poverty that prevents millions from purchasing adequate food even when it is available. Another reason is that in many cases starvation reflects not the absence of food but rather poor distribution due to politics or civil war, as in Sudan and Ethiopia during the 1980s. Societal excuses notwithstanding, in some parts of the world, the ability to provide food increasingly fails to keep pace with population growth. In sub-Saharan Africa, for example, the rate of population growth is 40 percent faster than the rate of growth in food production.

Scientists, economists, and philosophers have been fascinated by the notion of carrying capacity since Thomas Malthus suggested in 1812 that rates of increases in food production lagged so far behind population growth that starvation was inevitable. After years of study and debate, the definition and applications of this concept are still evolving. Malthus thought that ultimately a shortage of food would be the limiting factor on population growth, but his predictions did not take into account the remarkable advances in our ability to produce food.

THE GLOBAL HARVEST

Taking the world as a whole, Lester Brown, in the Worldwatch Institute's annual State of the World assessments, reports that grain production per person has climbed an impressive 40 percent between 1950 and 1984. The "green revolution" of the 1960s—which introduced new varieties of rice and wheat and intensified use of pesticides, fertilizer, and irrigation—is responsible for a major share of this increase. But from 1984 through 1988, grain production per person fell. While per capita grain production rebounded in 1989, it was still below 1984 levels. Such fluctuations do not suggest long-term trends or imply that environmental deterioration, rather than economic forces or isolated years of adverse weather, is solely responsible. They do illustrate that the earth's ability to supply food to the growing population cannot be taken for granted.

Discussions of how many people the planet can support often begin with some widely accepted numbers. Sometime in 1987, for instance, the world's population crossed the 5-billion mark. Demographers project that by the end of the coming century our numbers will increase to 10 billion or more before they begin to stabilize, with more than 95 percent of the growth occurring in the developing world.

Paul Ehrlich, Gretchen Daily, Anne Ehrlich, and Peter Vitousek, all at Stanford University, and Pamela Matson, at NASA Ames Research Center, are part of the Stanford Carrying Capacity Group. They attribute carrying capacity not only to the earth's physical and biological capabilities to provide resources necessary for food, clothing, and other essentials, but to humanity's ability to develop new technologies and attitudes as well. Through cultural evolution, they explain, human beings may quickly shift their demand for and ability to extract different resources. At the same time, natural and human-induced changes—with global environmental change as a primary example—alter the distribution and abundance of available resources. Food draws particular attention because its production is sensitive to changes in the environment, particularly to changes in temperature and precipitation, and yet basic human nutritional requirements are relatively inflexible.

The ability of the earth to produce food depends heavily on elements of what can be considered our species' capital: groundwater, the genetic diversity of nonhuman species, and productive soil. These natural assets, which population ecologist Ehrlich describes as part of "humanity's one-time inheritance," once seemed limitless. Now, he explains, groundwater in many places is being pumped faster than it is being recharged. Untold species and millions of genetically distinct populations that potentially provide the genetic resources for new crops disappear each year as tropical forests are cleared. Fertile soil, which is generated at rates on the order of inches per millennium, is losing its productivity in many parts of the world because of erosion or salinization, a process in which salts remain as irrigation water evaporates from the soil surface. Such direct human impact on carrying capacity is especially evident on marginal land in arid and semiarid regions, particularly in Africa.

A 1988 study by Robert Chen and colleagues for the Alan Shawn Feinstein World Hunger Program at Brown University estimated that even if food were equitably distributed (with nothing diverted to livestock), the amount of food produced in 1985—an all-time record—could have provided a minimal vegetarian diet to about 6 billion people, a number we will exceed by the end of the century. The same global harvest, allowing a diet with about 15 percent of the calories from animal products,

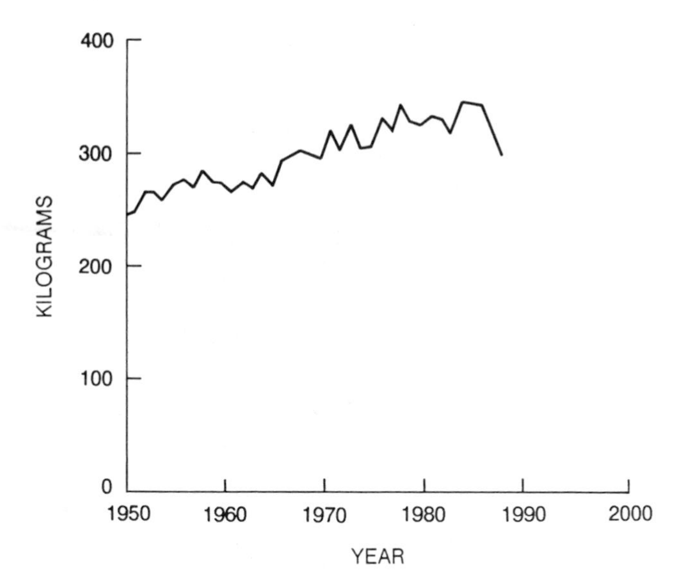

World grain production per capita, 1950–1988. (Reprinted, by permission, from *State of the World 1989*. Copyright © 1989, by Worldwatch Institute, all rights reserved.)

could feed some 4 billion people. A diet consisting of 35 percent animal products, similar to that consumed by most North Americans and West Europeans today, could be provided to only about 2.5 billion people—less than half of today's population. These estimates assume that 40 percent of the food harvested will not be available to human use because of wastage and consumption by pests. But as many economists have pointed out, if the poorest people had more wealth, the increased demand might well stimulate production of more food. So the issue of whether there would be enough food is never quite resolved.

Based on current projections for population growth and increases in per capita income, world demand for food in the middle of the next century could easily be 2 to 2.5 times the level of the mid-1980s, according to calculations by William Easterling III and Pierre Crosson, both of Resources for the Future, and Martin Parry, of the Atmospheric Impact Research Group at the University of Birmingham in the United Kingdom. Crosson and his colleague Norman Rosenberg, also at Resources for the

Future, believe there is room for optimism. They predict that if food production grows at or even slower than the current rate, there would still be enough food for the 10 billion people by the time they arrive. They temper this conclusion with the caveat that the ability to produce enough food can be sustained only if techniques that are less environmentally damaging than the current monoculture crops and heavy applications of chemical fertilizers and pesticides are developed and used.

The expansion of food production to feed the world's growing population is not likely to be accomplished by increasing the world's cultivated area. Even though only half of the potentially arable land is being farmed, expansion onto new land is limited because remaining land may be geographically inaccessible (uncultivated land is most scarce in many developing countries where population is growing fastest), infested by pests that transmit parasitic diseases such as trypanosomiasis (sleeping sickness), or covered by soils so thin or acidic that agriculture cannot be sustained. In fact, the land area planted in grain worldwide has actually declined by about 7 percent since 1981, owing mainly to abandonment of deteriorated land, conversion of cropland to nonfarm uses, especially in densely populated regions, and "set-asides" in the United States (a practice in which farmers are compensated for retiring cropland to limit overproduction). The primary prospects for expanding food production are the potential for increasing yields on existing agricultural land through more intensive cropping, increased fertilizer use, and the development of more productive strains, and increased reliance on and development of new methods for harvesting food from the oceans.

THE EFFECT OF GLOBAL WARMING ON FOOD PRODUCTION

Whether the projected global warming will be good or bad for agriculture depends on the specific location and on how much warming occurs. Little is certain as scientists try to sort out complexities such as how changes in temperature and precipitation patterns may affect agricultural productivity and also how crop yield may change as plants respond to increased concentrations of carbon dioxide, which can stimulate growth and reduce water consumption. Crop yields are certain to decline in some regions and increase in others. But researchers, lacking knowledge about the regional distribution of climate changes, can only estimate where and by how much these shifts in productivity will occur.

Scientists trying to simulate potential effects of global warming on agriculture rely on the same general circulation climate models (GCMs) that other scientists use to study global processes (see chapter 6). For the purposes of the agricultural models, the GCMs project some important changes in climate. In particular they find that warming will be greatest in the high latitudes, that soils may tend to be drier in mid-continental regions in summer, and that globally the hydrologic cycle will intensify—more rain, more evaporation—as the earth's surface warms. The models show that with an effective doubling of preindustrial carbon dioxide concentrations (that is, with a combination of all trace greenhouse gases that equals the heat-trapping effect of a doubling of the concentration of carbon dioxide), evaporation on a global basis will increase by 7 to 12 percent. The atmosphere cannot store large amounts of water vapor, and so precipitation will increase. The increases will not be uniformly distributed, however; nor will the proportions of rain, snow, or dew necessarily remain the same.

In a summary of the causes, impacts, and uncertainties associated with the greenhouse effect, Stephen H. Schneider, of the National Center for Atmospheric Research in Boulder, Colorado, and Rosenberg suggest that if analyses of the effects of temperature changes on evaporation and runoff of water from the land surface are correct, "the greatest impact of greenhouse warming on natural resources will occur because of changes in the seasonality and amounts of precipitation and of evapotranspiration."

Parry, with colleagues Timothy Carter, also of the University of Birmingham, and Nicolaas Konijn, of Agricultural University in the Netherlands, synthesized results of a multinational study of the impacts of climatic variations on agriculture. The study, sponsored by the International Institute for Applied Systems Analysis and the U.N. Environment Programme, used the Goddard Institute for Space Science (GISS) climate model to study the effects on crop yields of warming due to an effective doubling of carbon dioxide.

In the study, Parry and colleagues compared average yields expected under current climate conditions in several Northern Hemisphere regions with average yields that might be expected with effective doubling of carbon dioxide warming. For simplicity, they assumed the same technology and management used today, which they acknowledge is not a realistic assumption. They did not consider the fertilizing and moisture-saving effects of added carbon dioxide on the plants.

Parry, Carter, and Konijn report that if summer dryness becomes more frequent in mid-latitudes as predicted for the Northern Hemisphere, decreases in yields might occur in the productive areas of North America and the USSR. In general, they suggest, the higher temperatures would favor higher yields of cereal crops now grown in regions where current temperatures limit the growing season. For climate conditions produced by effective carbon dioxide doubling as projected by the GISS model, for instance, wheat yields increase by about one third in the central European region of the Soviet Union, where there is currently a short, cool growing season. The yields of barley, on the other hand, which thrives under cool temperate conditions, drop by about 4 percent. Where cereal production is already prone to drought, increased evaporation rates predicted by the climate model could limit any increase in yields that would be expected due to higher temperatures. This could be the case in Saskatchewan, for instance, where increases in yields of spring-sown wheat plants could be lessened by one fifth to one third.

The direct effects of carbon dioxide on plant growth and use of water complicate efforts to predict how future climate change induced by rising concentrations of greenhouse gases may affect agriculture, forests, and other ecosystems. As carbon dioxide concentrations increase, rates of photosynthesis increase in most plants. At the same time, with rising concentrations of carbon dioxide plants partially close their stomates, the pores through which water vapor and carbon dioxide pass. Because plants use less water (transpiration) per unit leaf area when exposed to elevated levels of carbon dioxide, water use efficiency may increase. So far, effects of carbon dioxide enrichment have been studied primarily under controlled experimental conditions. If the positive direct effects occur in the field, the combinations of increased growth and improved water use efficiency may help offset the negative effects of climate change on crops.

WATER SUPPLY, IRRIGATION, AND THE HYDROLOGIC CYCLE

One of the more generally accepted conclusions of the general circulation climate models is that as average global temperatures increase, the hydrologic cycle will speed up, increasing global precipitation. This does not mean, however, that the added precipitation will fall where or when it is needed. As temperature and precipitation patterns change, so will soil moisture and the timing and magnitude of runoff, with possibly adverse effects for many of the world's important agricultural areas. One likely consequence of these changes would be that demand for water, especially for irrigation, would increase in some regions.

The task of estimating future changes in water supply is difficult because the resolution of global climate models is too coarse to represent the complexity of regional or local changes. Many water problems such as floods and drought occur on these spatial scales. Despite their imperfections, however, the models tend to agree on several key points.

One point on which models tend to be in agreement involves changes in soil wetness, which may be just as important as changes in atmospheric temperature. The soil moisture regime determines the types and extent of vegetation that can thrive in a given location. Some studies, such as the ones conducted by Syukuro Manabe and Richard Wetherald, both of the NOAA Geophysical Fluid Dynamics Laboratory in Princeton, New Jersey, predict that as the concentrations of greenhouse gases increase, soil will become dryer in summer over vast expanses of the middle and high latitudes—including the U.S. Great Plains, Western Europe, northern Canada, and Siberia.

Model results suggest that there would be significant changes in runoff patterns under a changing climate. Runoff is sensitive to changes in precipitation and to evaporation, which is strongly affected by temperature. In many regions of the world, runoff comes as snow melts. With higher temperatures, relative amounts of rain and snow are likely to shift, as will the timing and speed of snowmelt. Peter H. Gleick, of the Pacific Institute for Studies in Development, the Environment, and Security in Berkeley, California, identifies in the 1989 publication *Greenhouse Warming: Abatement and Adaptation* a seasonal effect for basins in the western United States in which changes in runoff patterns may alter the likelihood of flooding and the availability of water during such times as the peak irrigation season. Similar changes are predicted for China, Canada, and Europe.

One implication of such findings is that if global warming becomes a reality, crop irrigation requirements are certain to increase, but farmers may find it difficult to expand the area of irrigated cropland, or even to maintain present irrigation levels. Agriculture already accounts for three quarters of the fresh water used globally. A 1989 National Research Council study states that in the United States, agriculture accounts for 85 percent of all consumptive uses of water; of this amount, 94 percent is used for irrigation. But if water supplies diminish, other uses such as industry, drinking water, and sanitation would also compete for the available fresh water.

Dean F. Peterson and Andrew Keller, both then at Utah State University, computed how three levels of climate change—3°C (5.5°F) warming, 3°C warming with a 10 percent increase in precipitation, and 3°C warming with a 10 percent decrease in precipitation—would affect irrigation requirements. In all three scenarios, irrigation increased because of the longer growing season, shifts in crops and more multiple cropping (more than one crop grown in a growing season), and greater potential evapotranspiration.

Irrigation is the underpinning of the world food production system. For millennia, farmers have relied on irrigation to increase yields of crops and to free them from the uncertainties in the timing and amount of rainfall. In some areas, irrigation makes farming possible; in others, it augments rainfall, with often dramatic results. In the United States, only 13 percent of the cropland is irrigated, but this land accounts for nearly one third of the value of crops produced. In 1985, the 270 million hectares (667 million acres) of the world's irrigated cropland provided nearly one third of the harvest. These predicted changes in precipitation and runoff would affect the availability of water for irrigation and hence food production.

EXPLORING AVENUES FOR ADAPTATION

The history of civilization is punctuated by swings of climate that have tried the ingenuity of people drawing their livelihood from the land. Sometimes societies can take steps to moderate the severe effects of climate swings by changing pricing structures or providing assistance to farmers.

In other cases, historians believe that climate change, combined with a lack of adaptation on the part of society, has led to the decline of civilizations. In a joint publication of the U.S. EPA and the U.N. Environment Programme, Martin Parry cites accounts of the Norse settlers in Greenland during the period between 1250 and 1500, the beginning of the Little Ice Age. The Norse settled along the Greenland coast around 985. By the thirteenth century the 6000 Norse inhabitants in two settlements faced a constellation of stressful circumstances: hostile Inuit, a declining European market for walrus ivory, and sequences of cool summers and stormy winters. The Norse did not opt to exploit the seas, as the Inuit did with such success, but continued to raise livestock despite the reduced capacity of the pastures.

Between 1350 and 1450, the Norse abandoned the settlements, while the Inuit continued to survive there.

The experience of the Norse settlers is an extreme example, Parry explains, of how societies "can fail to identify and implement appropriate policies of response, not only to climatic change but to the synergistic effects of a number of concurrent events." This historical lesson, he suggests, shows the value of designing policies that respond to the host of difficult environmental problems facing us today. The effects of climatic disruption can reverberate throughout an entire society, from the fortunes of specific farmers and regions, to national and global food supplies, to trade imbalances and the global economy.

As the initial steps in analyzing the impacts of climate change on agricultural yields, the studies mentioned above largely assume that farm policy, management, and technology remain as they are today. But these systems are far from static. It is inconceivable that farmers will not react. Their livelihood is defined by continual adjustment to changing patterns of weather and consumer demand. Farmers in the midwestern United States, for instance, after experiencing the 1988 drought that reduced corn harvests nationwide by almost 40 percent, took special care to sow their 1989 spring crops early in case the drought persisted. Besides adjustments in planting and harvest dates, other important adaptations at the farm level include changes in tillage practices; crop varieties, species, and rotations; and fertilizer, herbicide, and pesticide applications. Farmers may also improve existing irrigation efficiency or install new irrigation facilities. At the regional level, agricultural market, transportation, finance, and water resources infrastructures are very likely to change, as are national farm policy and international trade agreements.

An analysis by Easterling, Parry, and Crosson finds that if growing seasons in cool regions become longer and warmer, farmers could increase yields substantially by substituting varieties used today for varieties that already thrive elsewhere under higher temperatures. Under conditions of effective carbon dioxide doubling, if late-maturing rice now grown in central

Japan, for instance, were planted in northern Japan, yields might increase by 26 percent.

The efficacy and the cost of these adjustments depend in part on the severity of the climate change experienced. As Cynthia Rosenzweig, a researcher at Columbia University and NASA's Goddard Institute for Space Studies, points out in testimony presented to the U.S. House of Representatives Committee on Agriculture, "Farmers can make some adjustments to less severe climate change by planting earlier, substituting better-adapted crop varieties and species, and increasing demand for water for irrigation. More severe climate change will likely require major adaptations, including expansion of irrigation infrastructure, farm abandonment and rural dislocation, in some regions."

A special burden may fall on the half billion poorest and hungriest farmers of the world. Robert Kates, director of the Alan Shawn Feinstein World Hunger Program at Brown University, notes that they are increasingly finding themselves restricted to ecologically marginal land and water resources as their numbers increase and traditional access to important seasonal uses of land or water are lost to development, dams for electricity production, large farms for export crops, or even wildlife and forest conservation. Forced onto marginal land, they add to its degradation. And they may be the ultimate victims of global change, having neither the resources to take advantage of climates more favorable for agriculture nor the resources to cope with a less productive climate.

Although critical uncertainties exist about the magnitude and timing of predicted warming and agricultural systems are sure to adjust in many ways, climate change raises long-term concerns about agricultural productivity, depletions of major resources (especially land and water), viability of rural communities, and the environment. The concurrent projections of population increases and vulnerability of carrying capacity described early in this chapter can only add to the already enormous challenges currently facing global agriculture.

8

Coastlines and Rising Seas

Rising sea level may be one of the most widely felt and easily recognized consequences of a warmer global climate, but the inundation it would bring is already familiar to the millions of people who live and work in the world's coastal areas. Over the centuries, engineers have struggled to protect low-lying coasts from advancing waters. Their techniques range from straightforward measures such as pumping sand onto eroded beaches, building levees, and relocating ports, to complex, multibillion dollar systems of carefully maintained dikes and sand dunes, such as the one that protects more than half of The Netherlands from inundation by the North Sea. If the earth's surface temperature grows warmer, as many researchers predict, the task of keeping the sea at bay will become more difficult, and more urgent.

Scientists expect global sea level to rise for two reasons. First, as greenhouse gases accumulate in the atmosphere and eventually raise the earth's surface temperature, glaciers and land ice around the world will melt more rapidly, releasing water that will raise average sea level. Second, as the ocean absorbs additional heat from the air above it, the water will expand.

Like climate, sea levels have fluctuated throughout the earth's history. Eighteen thousand years ago during the last ice age, mean sea level was 100 to 150 meters (about 330 to 500 feet) lower than it is now. About 100,000 years ago during a period when the global average temperature was 1° to 2°C (about 2° to 3.5°F) warmer than today and the glaciers receded, sea level was about 5 to 7 meters higher. Even during the past 6000 years, sea level in some places has varied by a meter or more.

Currently, sea level is rising. Scientists estimate that in this century, sea level has risen by about 0.15 meter or more. Though it is tempting to correlate this rise with the coincident buildup in atmospheric carbon dioxide, a cause-and-effect relationship has not been established. The reason for the rise is poorly understood and exceedingly difficult to determine. Mark Meier, a glaciologist at the University of Colorado in Boulder, cautions that any projections for future sea level rise must be "tempered by the uncertainty of our understanding of the present."

Estimates made in the last decade projected that we could expect global average sea level to rise from about 0.5 to 1.5 meters, with some estimates as high as 3 meters, in response to a 3° to 5°C warming that some estimate would be expected by 2050 with an effective doubling of carbon dioxide. Recent estimates have revised this figure downward to about a third of a meter, primarily because of new calculations on how snowfall will increase on the polar ice sheets (warmer air can transmit more moisture, resulting in more snowfall) and thus bind up the water that would otherwise go into the oceans and on how long it will take meltwater to escape from subfreezing glaciers. Scientists are still working to resolve the many uncertainties surrounding these projections.

The earlier high estimates considered the possibility that the unstable West Antarctic Ice Sheet might disintegrate, releasing about 2 million cubic kilometers of ice into the ocean. Scientists increasingly believe that such a collapse, if it occurs, would be slow and that little effect would be seen during the coming century. On the other hand, as concentrations of greenhouse gases build up past the effective carbon dioxide doubling benchmark, researchers believe that global surface temperature will continue to increase. As it does, Meier explains, "Our feeling is that sea level rise will continue. In fact, it will accelerate. The problem is that we don't yet know how to estimate very well what is going to happen."

About some processes and conditions related to sea level rise, scientists are confident. Worldwide average sea level depends simply on the volume of water resting in the ocean basins. Locally, at a particular coast or island, sea level is influenced by the characteristics of the locale, including its geology and the way humans manage the land. For this reason, global sea level may rise a little in some places and a lot in others. In some places it may even drop. The human and economic costs also would vary from nation to nation.

Some parts of the world, such as Canada, Scandinavia, and Scotland, are rising, springing up in response to the removal of glacial ice that began with the end of the last ice age. In such places rates of uplift exceed past rates of sea level rise, so that the sea level appears to be falling. In other areas, such as Bangkok and most of the U.S. Atlantic coast, sea level appears to be rising more rapidly than the global, or eustatic, change. In fact, the land itself is subsiding in response to downwarping of the earth's crust from slow geologic processes, sudden earthquakes, or compaction of soft underlying sediments such as mud or peat. Human activities such as pumping of groundwater, oil, and gas from below the land surface can be major factors in subsidence. Because sea level in any one location is subject to these many factors, it is measured in relative terms, namely the difference between the eustatic change and any local change in land elevation.

THE HUMAN COSTS OF RISING SEA LEVEL

As many as one billion people, or 20 percent of the world's population, live on lands likely to be inundated or dramatically changed by rising waters. Many buildings are at elevations low enough that even at current sea levels, the safety of people during storms cannot be assumed. On the basis of current predictions of sea level rise, a 1989 Resources for the Future report explains that all land currently up to 5 meters (about 16.5 feet) above mean sea level is potentially vulnerable in the next one to three centuries to upstream effects of flooding, particularly storm surges—walls of water pushed by heavy winds—and to intrusion of saltwater into groundwater and estuaries.

The effects of sea level rise will vary with conditions along the coastlines. Sandy areas and unconsolidated cliffs of sedimentary material will erode quickly, whereas cliffs of resistant rock will not. Steep-sloped areas will be affected by each increment of rising water but less so than gently sloping coastal terrain, which would be flooded by the same sea level rise. An extreme example is the Republic of Maldives, a nation of 1190 small islands, most of which are barely higher than 2 meters in elevation. If mean sea level rises 2 meters, the entire country would be submerged. Even if mean sea level rises 1 meter, a storm surge would be, as Maldives president Maumoon Abdul Gayoom told the U.N. General Assembly in 1987, "catastrophic and possibly fatal to the nation."

Some of the most visible effects of predicted changes in climate will be along the ocean coastlines, where rising seas would flood wetlands and lowlands and hasten coastal erosion. Water tables near the coast would rise, and saltwater would intrude into rivers, bays, and the underground aquifers that supply water for drinking, industry, and irrigation. In developed areas, buildings, roads, and sewage systems would be threatened. Gjerrit P. Hekstra, of the Ministry of Housing, Physical Planning, and Environment in The Netherlands, reports that if maximum storm surges and the upstream effects of flooding are taken into account, about 3 percent of the land area worldwide would be subject to inundation or made vulnerable by intrusion of saltwater. This area includes one third of the world's cropland because productive soils commonly occur in delta areas. Estuaries, where freshwater rivers flow into the sea

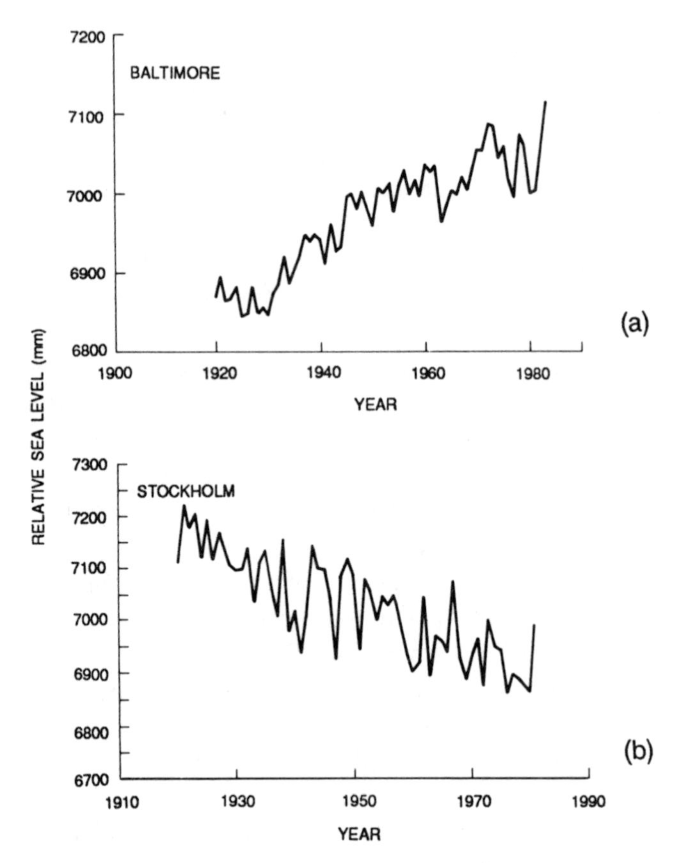

Tide-gauge records from the early twentieth century indicating (a) rising local relative sea level at Baltimore on the North American East Coast and (b) falling local relative sea level at Stockholm in Scandinavia, as the land rises in "rebound" from the weight of receding glaciers. (Courtesy D. G. Aubrey and A. R. Solow, Woods Hole Oceanographic Institution.)

and mix with the incoming tides, and groundwater aquifers are most vulnerable to intrusion. According to James Titus, of the EPA, increased salinity already is considered a cause of reduced oyster harvests in the Chesapeake Bay and has hastened the conversion of cypress swamps in Louisiana to open water.

Perhaps no other impact of climate change would displace more people from their homes than rising seas, swelling a growing class of "environmental refugees." Not surprisingly, most of these refugees would be in developing countries, where scarce economic resources cannot accommodate the costly remedies required to keep the sea at bay.

The system of wetlands that grace many stretches of the world's coastlines may be a prominent victim of sea level rise. Coastal wetlands include marshes, swamps, and salt-tolerant mangrove swamps in the tropics. Wetlands are nurseries for many terrestrial animals and commercial fish species, and the ecosystems they support are among the most productive on earth. They help maintain water quality and protect the shorelines from erosion and the full brunt of coastal storms.

For centuries people have drained and filled wetlands, converting them to cropland or urban areas. Where navigational channels transect wetlands, or where flood control structures have been installed, wetlands have been converted to open water. In the 1960s, however, environmental awareness began to increase, and many governments have taken steps to curtail wetlands destruction. Titus explains that in the United States these restrictions have substantially reduced the conversion of coastal wetlands to drylands but that the conversion of wetlands to open water continues. Louisiana, for example, loses 50 square miles per year to open water.

How the wetlands respond to sea level rise will depend in part on the degree to which they have been developed. To protect their property from the sea, people in many areas have built bulkheads just above the wetlands. Many coastal ecologists fear that these obstacles squeeze the wetlands against the advancing sea.

In the absence of human activity, coastal ecosystems are preserved by a highly effective mechanism that helps compensate for fluctuating sea levels. The rising waters flood areas that are now dry land, creating new wetlands, which can grow upward as sediment and organic material accumulate. Where coastlines are heavily populated and developed, this mechanism may have only limited potential to prevent a major loss of wetlands in the coming century.

Titus cautions that the area of wetlands today is far greater than the area that would be available for new wetlands as sea level rises. In its report to Congress on the *Potential Effects of Global Climate Change on the United States*, the EPA reports that a 1-meter sea level rise by 2100 could drown 25 to 80 percent of U.S. coastal wetlands. Their survival depends on whether they can migrate inland, or if levees and bulkheads built to protect coastal properties block their path.

Nor is the future bright for the string of barrier islands that adorns the eastern United States and Gulf coasts. Barrier islands such as Hatteras Island, North Carolina, and Long Beach Island, New Jersey, are long, narrow islands and peninsulas only a few feet above sea level in elevation, with one side facing the ocean and the other side facing a bay that separates the island from the mainland. Left on their own, barrier islands keep pace with sea level rise by migrating, rolling landward as storms wash sand from the oceanside beach to the bayside marsh. Even without the obstacles presented by human-built structures, barrier islands may disintegrate as sea level rise speeds up. Coastal Louisiana, for instance, is subsiding so rapidly that relative sea level is rising by 1 centimeter (about 0.4 inch) per year, and its barrier islands are breaking up. (This already serious situation will worsen if the rate of sea level rise accelerates in the next century.)

While rising waters may inundate the baysides of the barrier islands, the oceansides of the islands and the mainland coastal beaches are subject to scouring wave action and are particularly at risk for runaway erosion. Surveys indicate that during the past century, less than 10 percent of the length of the world's sandy shorelines advanced seaward, while more than 60 percent retreated. In the United States, a 1-foot rise in sea level could cause beaches from the Northeast to the mid-Atlantic to erode 50 to 100 feet. Along the Carolinas, the sea would advance 200 feet; in Florida, where shoreline protection already costs millions yearly, the shoreline would retreat from 100 to 1000 feet; and in Louisiana, several miles. EPA notes that at high tide, most U.S. recreational beaches are less than 100 feet wide.

THE PLIGHT OF THE DELTAS

A rise in global sea level would be most deeply felt in river deltas, the wetlands that form when a river carries more sediment into a body of water than can be carried away by currents and waves. This sediment builds up to form the delta, which acts as a barrier between the land and the sea. Under natural conditions, deltas form and break down continuously as they accrete and subside. But when humans intervene by damming, diverting, or channeling rivers, the amount of sediment reaching the delta is reduced. The delta is thrown out of equilibrium, with the accumulation of sediment no longer offsetting the subsidence.

The Mississippi River delta in Louisiana is the most vulnerable area to relative sea level rise in the continental United States and provides a ready example of the changes under way in many of the world's deltas. The fine-grained sediment carried by the river to the sea tends to compact and subside in the manner typical of deltas. Normally this effect is counteracted by annual flooding, during which the Mississippi overflows its banks and supplies sediment to the wetlands. But measures to trap sediment upstream in order to reduce dredging costs at the mouth of the river, and levees built for flood control, prevent sediment, fresh water, and nutrients from reaching the wetlands. Titus estimates that if current management practices continue and sea level rises as projected, most of Louisiana's wetlands will be destroyed in the next century.

Low-lying countries in the developing world such as Egypt and Bangladesh, where rivers are large and the deltas extensive and densely populated, will be hardest hit by sea level rise. Where the rivers are dammed, as in the case of the Nile, the effects of inundation and coastal erosion will be especially severe. John Milliman, James Broadus, and colleagues at the Woods Hole Oceanographic Institution studied the economic and human consequences of sea level rise for Egypt and Bangladesh. They estimate that the combined effects of sea level rise and subsidence caused by extraction of groundwater could threaten the

"Lamentable News out of Monmouthshire": Old woodblock illustration of fourteenth-century coastal flooding in England. (Courtesy of University of East Anglia, Norwich, United Kingdom.)

homes and livelihoods of 46 million people in these two countries alone. In their analysis, the researchers used two estimates for sea level rise: a minimum of 13 centimeters by 2050 and 28 centimeters by 2100, and a maximum rise of 79 centimeters by 2050 and 217 centimeters by 2100. They then developed three possible scenarios. The "best case" assumes that the minimum rise in global sea level occurs and that the delta is in equilibrium. The "worst case" assumes the maximum rise in sea level and the complete damming or diversion of the sediment carried by the river draining into the delta (in this case the sediment would not be able to replenish the delta so that natural subsidence would increase the relative rate of sea level rise). The "really worst case" assumes that the effects of natural subsidence are compounded by excessive removal of groundwater as well as by the maximum increase in global sea level.

Almost all of Bangladesh lies in the massive Bengal delta formed where the Ganges, Brahmaputra, and Meghna rivers converge after traversing India, Nepal, Bhutan, Tibet, and China.

Half of the country is at elevations of less than 5 meters. The Woods Hole researchers find that in the best case, sea level would rise about 13 centimeters by 2050 and less than 1 percent of the nation's total land area would be lost to advancing seas. In the worst case, 18 percent of the land, which currently supports about 15 percent of the nation's people, would be lost. In the really worst case, by 2100 waters would cover an area currently supporting 35 percent of the nation's population. The land area that supports nearly one third of the nation's current gross national product would vanish into the sea.

Global warming could further compound the miseries of Bangladesh if the hypothesized possibility of more frequent storms occurs. Currently, an average of one and one half severe cyclones hit Bangladesh each year, and the storm surges advance up the rivers as far as 200 kilometers inland. If the storms increase in frequency and sea level rises, the storm surges will reach even further toward the main centers of population.

Egypt too would experience massive change with global sea level rise. Though Egypt is much larger than Bangladesh, its people live on only about 3.5 percent of the land. Within this area, population density—1800 people for every square kilometer—is twice that of Bangladesh. Much of Egypt's productive agricultural land is within the Nile delta.

The Nile River has been completely dammed since completion of the High Dam at Aswan in 1964. For this reason the Nile delivers no sediment and little fresh water to the Mediterranean, where the delta fans out into the sea. This loss of sediment combines with subsidence to cause dramatic erosion of the delta shores. Between 1966 and 1974, many areas eroded more than 1 meter per year, and some locations lost more than 100 meters per year.

Egypt is precluded from the best case scenario because the Aswan Dam already traps the sediment. The Woods Hole researchers report that the worst and really worst cases are more likely. In the worst case, local sea level would rise between 13 and 133 centimeters by 2050, which would affect as much as 19 percent of the currently habitable land. In the really worst case, sea level would rise between 28 and 332 centimeters by 2100 and would affect up to 26 percent of the habitable land. Several brackish lakes behind the coastal dunes supply approximately half of the nation's fish catch and would be endangered.

Other areas with low-lying deltas and large rivers could experience similar fates. The researchers suggest that although there may be little that individual developing countries can do to control global climate change, they can control groundwater extraction and the trapping of sediments in dams. These factors are perhaps more responsible than global climate change for local changes in relative sea level.

PREPARING FOR RISING SEAS

In many cases, decisions about how to prepare for rising sea levels come down to stark economic facts that will vary with specific countries and localities. A 1987 report from the National Research Council on the engineering implications of sea level rise suggests that "holding back the sea as water levels rise will almost always be feasible," but that in some cases it may not be economically or environmentally sound. Responses fall into three main categories. Humans can retreat from the shoreline, use structures that prevent flooding and shoreline recession, or raise the land.

For some cities the choice whether to protect is obvious: Manhattan will never surrender to rising seas. In fact, most well-developed coastal communities will probably decide that protective measures such as bulkheads, levees, and pumping systems, and the high costs of maintaining them, are worthwhile, whereas sparsely developed areas may continue to adapt as the shoreline retreats. The EPA estimates that low-lying coastal cities in the United States could be protected from even a 2-meter (about 6.6 feet) sea level rise. A conservative cost estimate ranges from $30 billion to $100 billion, but these sums would still be a small fraction of the values of the coastal cities. Cost estimates for raising the nation's recreational barrier islands, which cover a combined area of 100 to 150 square miles, range from $50 billion to $100 billion for a 1-meter rise and from $135 billion to $215 billion for a 2-meter rise. The trade-offs between protection from rising seas and economic costs of protection will be difficult to balance. In addition, levees and seawalls hold back the sea but also block waterfront views, cause the beach to erode, and prevent the migration and preservation of wetlands. The choice of response should be based on a sound understanding of coastal processes. If people opt to stem destruction of wetlands, they may have to surrender control of these areas to natural forces. This would mean removing artificial structures that thwart sedimentation and migration of wetlands, or moving them back from the shore as the waters advance. Coastal scientists, engineers, and policy analysts suggest that some communities may plan an orderly retreat from the shoreline rather than retreating in response to future rises. The National Research Council report cautions that because methods for dealing with coastal erosion are evolving rapidly, communities should periodically reconsider any plans to retreat.

The projected sea level rise should cause neither alarm nor complacency, researchers stress. The severity of impacts from sea level rise depends on how society responds and adapts. People are adept at using and advancing technology when necessary and appropriate.

So far, the projected acceleration in the rate of sea level rise has not been detected. If the rate does accelerate, the increase initially would be so slow that we might not detect it until the coming century. But some responses, especially those requiring construction of costly and massive structures, can take three to four decades from planning stages to completion. Some decisions must be made soon if defenses are to be ready in time.

No one can anticipate all the changes that a warmer climate will cause. Of the changes that are expected, rising sea level could have the most visible effects on society, literally forcing people to abandon their homes or to build virtual fortresses around their cities. Whereas the industrialized countries of the world that have contributed most heavily to the increase in greenhouse gases conceivably could afford the technological remedies to cope with rising waters (though it will cost them dearly), many developing nations could not. As Tom Goemans, a consultant with KPMG in The Hague, notes, accelerated sea level rise is a "manmade phenomenon that presents a perfect example of external costs, that is costs not included in the present price of energy from fossil fuel burning."

9

The Ozone Layer and Ultraviolet Radiation

The first unmistakable sign of human-induced change in the global environment arrived in 1985 when a team of British scientists published findings that stunned the world community of atmospheric chemists. Joseph Farman, of the British Meteorological Survey, and colleagues reported in the scientific journal *Nature* that concentrations of stratospheric ozone above Antarctica had plunged more than 40 percent from 1960s baseline levels during October, the first month of spring in the Southern Hemisphere, between 1977 and 1984.

Most scientists greeted the news with disbelief. Existing theory simply had not predicted it. It meant that for several months of the year a hole forms in the ozone layer, which protects animals and plants from ultraviolet solar radiation. Suddenly it seemed that the chemical processes known to deplete ozone high in the earth's atmosphere were working faster and more efficiently than predicted. The discovery brought home a critical fact about the planet. No matter how much we learn about the workings of the earth system, the unexpected can always occur.

Ground-based observations conducted by the National Oceanic and Atmospheric Administration (NOAA) since 1964 had not revealed the drop. Measurements taken by the Total Ozone Mapping Spectrometer aboard the Nimbus 7 satellite operated by the National Aeronautics and Space Administration (NASA) since 1978 reflected the change but had not yet been analyzed. When researchers scrutinized the data, Farman's findings were confirmed, but tough questions remained. What processes were causing the hole? Would the thinning of the ozone layer spread to other latitudes, or was it confined to the Antarctic?

To gather more information about antarctic ozone chemistry and the ozone hole and its causes, a team of scientists led by atmospheric chemist Susan Solomon, of NOAA in Boulder, Colorado, headed off in 1986 on the first National Ozone Expedition to the Antarctic. By 1987 they and other teams of researchers had learned that the ozone over Antarctica had been reduced by more than 50 percent of values recorded in 1979, the first October of satellite operation, and that at altitudes between 15 and 20 kilometers, depletion was as great as 95 percent. In 1988 temperatures (which influence processes in the stratosphere) were milder than in 1987, and in October the ozone declined by about 15 percent of 1979 values (already 20 percent below the baseline values of the 1960s). In 1989 temperatures dropped again, and ozone levels matched the severe depletion of 1987.

CHEMISTRY OF THE OZONE LAYER

Until the hole was discovered, scientists were fairly sure that they understood the chemical processes at work in the ozone layer. Oxygen molecules (O_2), abundant throughout the atmosphere, are split apart into individual atoms (O+O) when energized by radiation from the sun. These atoms are free to collide with other O_2 molecules to form ozone (O_3). The particular configuration of the ozone molecules allows them to absorb the sun's radiation in ultraviolet wavelengths that are harmful to life if they penetrate to the earth's surface.

The ozone molecules formed by collision are partially removed by other naturally occurring chemical reactions, and so

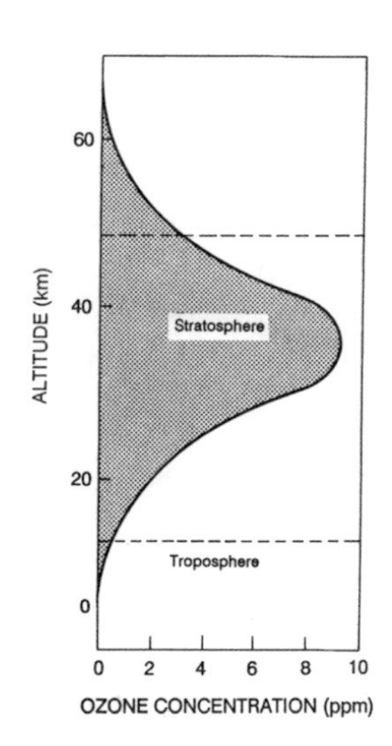

Ozone is found in varying concentrations from the earth's surface to a height of some 60 kilometers. Its concentration increases sharply in the stratosphere. Maximum ozone concentrations occur at a height of 25 to 35 kilometers but even here never exceed about 10 parts per million by volume. (Adapted from U.N. Environment Programme. 1987. *The Ozone Layer,* Fig. 2, p. 9. Copyright © 1987, U.N. Environment Programme.)

the overall concentration of stratospheric ozone remains constant. High above the stratosphere, the density of gases is so low that oxygen atoms rarely find other molecules to collide with, and ozone does not form in abundance. Below the ozone layer, too little solar radiation penetrates to allow appreciable amounts of ozone to form. Thus most of the world's ozone is in a stratospheric layer bulging with ozone at altitudes from 10 to 35 kilometers.

Closer to the ground, in the troposphere, ozone produced through a series of chemical reactions involving hydrocarbons and nitrogen oxide emissions from vehicles and industrial activity is an effective greenhouse gas (in addition to having adverse impacts on human health at high concentrations). Thus ozone plays two very different roles in global environmental change: one in the stratosphere as a shield against harmful ultraviolet radiation, and another nearer the ground in the troposphere as a greenhouse gas and a health hazard.

It is now known that in addition to the naturally occurring chemical reactions in the stratosphere, certain reactions involving chemical species of industrial origin, including chlorine and bromine, also chemically destroy ozone molecules. Atmospheric chemists F. Sherwood Rowland, of the University of California at Irvine, and Mario J. Molina, now at Massachusetts Institute of

Technology, first hypothesized the link between natural ozone in the stratosphere and chlorine released into the atmosphere from industrial sources. In 1973 they began to wonder: What happens to the industrially produced chlorinated molecules that are released into the lower atmosphere and for which no natural mechanisms for removal are known? The only long-lived natural source of chlorine in the earth's atmosphere is methyl chloride, which comes from the ocean and is present in the atmosphere at low levels.

The researchers hypothesized in 1974 that increasing concentrations of chlorofluorocarbons (CFCs), synthetic compounds that are chemically very stable in the lower atmosphere, rise unchanged through the lowest atmospheric layer, the troposphere. Even though CFCs are produced mostly in the industrialized countries of Europe and North America—where they are used in a wide variety of applications such as for solvents and refrigerants—they mix throughout the lower atmosphere, so that there are as many CFC molecules over Antarctica as over Colorado or Washington, D.C. The researchers surmised that upon reaching the stratosphere, the CFCs encounter high-energy ultraviolet light, which breaks them down, releasing their chlorine atoms. The chlorine atoms can then engage with ozone in a catalytic reaction in which each chlorine fragment can destroy up to 100,000 ozone molecules before other chemical processes remove the chlorine from the atmosphere.

The hypothesis was borne out and improved by measurements and observations. In 1970 chlorine was present in the stratosphere at 1.2 parts per billion, and at about 3 parts per billion in 1985. Were CFC use to continue at the 1985 rates (an eventuality precluded in 1987 by an international agreement known as the Montreal Protocol, described below), the stratosphere would contain about 8.2 parts per billion of total chlorine in the year 2050; current models of the chemistry and physics of the stratosphere suggest that at this concentration, total global ozone would drop by 5 percent.

Rowland and Molina believed that most of the chlorine molecules that reached the stratosphere would form relatively inactive and harmless compounds. The ozone depletion would occur gradually, they hypothesized, and might not be detected for many years. As ozone was lost, more ultraviolet radiation would reach the earth's surface. The researchers said two of the CFCs—CFC-11, which is widely used as a blowing agent in plastic foam, and CFC-12, mostly used as a refrigerant—were particularly likely to destroy ozone because of their widespread use.

These two CFCs alone are increasing in the atmosphere at an annual rate of about 5 percent. They are part of a class of chemicals known as halocarbons, many of which attack and destroy stratospheric ozone and also contribute to global warming as greenhouse gases. Another chlorofluorocarbon, CFC-113, is used as a solvent for cleaning electronic circuitry. Its atmospheric concentration is going up at an annual rate of about 11 percent. Scientists are beginning to eye concentrations of still other synthetic halocarbons with suspicion. These include carbon tetrachloride, which is used as a cleaning fluid and in CFC production; methyl chloroform, used in solvents and adhesives; and halon 1301 and halon 121, which are used in fire extinguishers. Bromine, a chemical element that is related to chlorine and which is released from compounds used in fumigants and some fire extinguishers, is accumulating rapidly in the atmosphere. Bromine is believed to cause 10 to 30 percent of the antarctic ozone depletion.

STUDYING THE ANTARCTIC OZONE HOLE

In the year before the discovery of the ozone hole, scientists were estimating that increasing use of chlorofluorocarbons might cause reductions in the total ozone at high latitudes by about one percent in the 1980s and by 5 to 10 percent 50 to 100 years from now. "While those numbers were disturbing," Solomon said, "they were nevertheless small enough that it was hard to argue that they were even real. They were not being observed."

That viewpoint changed—along with scientists' faith in their

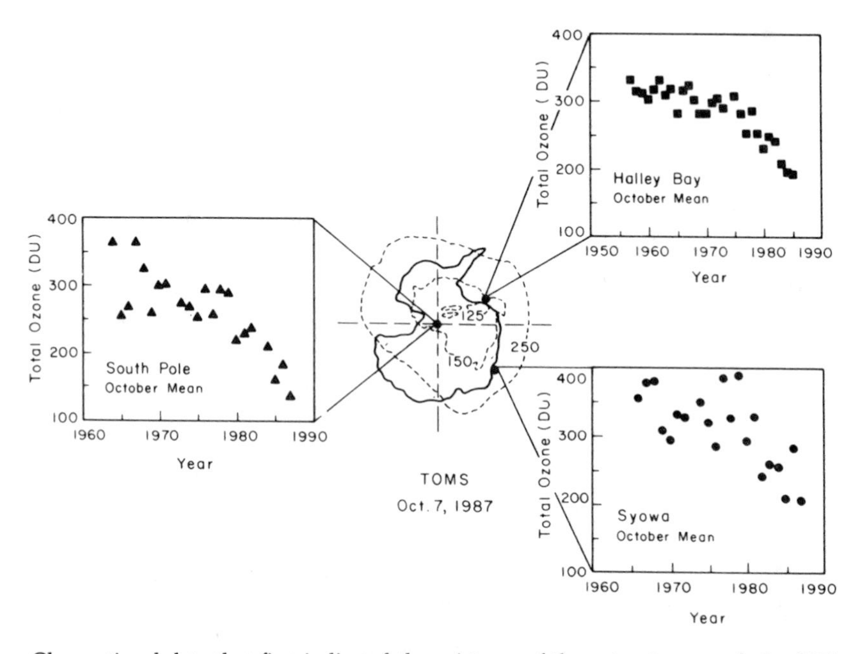

Observational data that first indicated the existence of the antarctic ozone hole. (DU, Dobson units; TOMS, total ozone mapping spectrometer.) (Reprinted, by permission, from *Global Change and Our Common Future.* Copyright © 1989, National Academy Press, Washington, D.C.)

models—in the mid-1980s as observations poured in from the coldest place on earth. Now, many scientists describe the antarctic ozone hole as the first unequivocal evidence of ozone loss due to man-made chlorine and one of the first clearly definable effects of human-induced global change.

When the antarctic ozone hole was first discovered, little was known about the antarctic stratosphere beyond the ozone measurements themselves. Virtually no data were available on the other chemical compounds present in the stratosphere, nor was there detailed meteorological information. This information was gathered rapidly by means of aircraft and state-of-the-art instrumentation. In short order, scientists were able to measure a broad range of atmospheric compounds, including chlorine monoxide, chlorine dioxide, hydrochloric and nitric acid, nitric oxide and nitrogen dioxide, and nitrous oxide. They found that the ozone levels dip at about the same latitudes where levels of chlorine monoxide ascend. As one researcher quipped, "These measurements are better than a smoking gun. This is more like seeing the guy pull the trigger." Scientists can now calculate how much ozone would be lost with a given amount of chlorine monoxide. The answer is strikingly similar to the levels of ozone depletion observed. Scientists are convinced that the elevated levels of chlorine and bromine account for much, if not all, of the antarctic ozone depletion.

For most of the year, the atmosphere over Antarctica has fairly high ozone concentrations. The ozone molecules are formed over the tropics and are delivered along with chlorine to the Antarctic, as well as to the Arctic, via atmospheric motions. In Antarctica, a circulation pattern known as the antarctic polar vortex traps the ozone over the South Pole for several months. It is within this vortex that scientists have measured such shockingly low ozone concentrations during the first two weeks of October, shortly after the beginning of the Southern Hemisphere spring.

The explanation for the decrease lies in the combination of ozone-destroying chemistry and weather conditions that favor formation of the high, thin clouds known as polar stratospheric clouds (PSCs). The stratosphere is extremely dry, and the ice crystals that make up the clouds form only when temperatures drop to −80°C (−112°F) or lower. The clouds foster a basic change in stratospheric chemistry by allowing reactions to occur on surfaces rather than between gas molecules. The chemical reactions that take place on these surfaces convert chlorine from forms that do not react with ozone to other, less stable forms that readily break up in the presence of sunlight and go on to destroy ozone. Both cold temperatures and sunlight are critical to the process leading to ozone depletion in the Antarctic. Antarctic ozone is depleted not during the winter, when temperatures are coldest and the South Pole is immersed in darkness, but in the southern spring, after sunlight returns but temperatures are still low.

Researchers describe a process something like this: Usually chlorine in the stratosphere becomes trapped in so-called reservoir compounds, such as hydrogen chloride and chlorine nitrate, which themselves do not destroy ozone. Once the stratosphere becomes cold enough that cloud particles freeze, the ice crystals provide surfaces on which reactions can occur: chlorine nitrate ($ClONO_2$) reacts with hydrochloric acid (HCl) present on the ice surface, producing molecular chlorine (Cl_2) and nitric acid (HNO_3). The nitric acid remains bound to the ice, and the molecular chlorine is quickly broken down into atomic chlorine (Cl). The chlorine atoms react with ozone (O_3), destroying it through the production of chlorine monoxide (ClO) and molecular oxygen (O_2). In a vicious cycle, the chlorine monoxide undergoes further reactions that re-form a chlorine atom, which is then free to destroy another ozone molecule.

As researchers improve their understanding of the antarctic ozone hole, it seems less ominous than it did at first for most of the rest of the world. Over the mid-latitudes in the Southern Hemisphere, however, the hole may be spreading. Recent research suggests that in the late spring, when the antarctic vortex breaks up, the winds transport the polar, ozone-depleted air into lower latitudes. The record low ozone values found over the Antarctic in October 1987 were followed by record low levels over Australia and New Zealand that December as the Southern Hemisphere summer began. NASA's Ozone Trends Panel reports that the effect may persist year round and that since 1979 ozone levels at all latitudes south of 60°S have decreased by 5 percent or more.

For the most part, the hole has not spread outside of Antarctica and the lower Southern Hemisphere because it is limited by the seasons and the frigid temperatures required for the formation of the ice-laden polar stratospheric clouds. Yet, the insights gained during several years of intense data-gathering have raised concern about ozone in the stratosphere over the rest of the globe.

OZONE DEPLETION IN OTHER LATITUDES

With ozone levels over the South Pole dropping up to 50 percent or more for several months each year, scientists are eager to know whether the same processes are operating to deplete ozone over the Arctic. Results gathered by scores of atmospheric scientists using sensors aboard airplanes and balloons suggest that the arctic stratosphere differs from the antarctic stratosphere in a number of important ways that make a northern ozone hole of the same magnitude unlikely.

Measurements from satellites and ground-based stations reveal ozone losses of about 5 to 10 percent at northern high latitudes during the arctic winter. This is much smaller than in the Antarctic for several reasons. For one, the arctic stratosphere generally warms up much earlier in the spring than does the antarctic, and the average temperatures are warmer. This means that cold temperatures and the sunlight necessary for the formation of polar stratospheric clouds—and the ozone depletion they promote—overlap for a much shorter interval. Another factor is that the arctic vortex is not as tight as the antarctic vortex. As Rowland explains, air drifts across the pole, through the polar darkness, undergoes some polar stratospheric cloud chemistry, emerges into sunlight—still in arctic winter—and loses a little ozone. Then the vortex warms up, and the ozone loss in the air mass stops. Meanwhile, another air mass is coming, and the process of successive small losses is repeated throughout the winter.

So far, the timing of the warming in the Arctic has offered some protection against wholesale ozone depletion. But researchers worry that this may not always be the case. In the winter of 1988–1989, the arctic winter was unusually cold—the coldest for at least 25 years. In January 1989 the polar vortex was relatively stable, giving rise to conditions similar to those in the antarctic winter stratosphere. In late January near the Arctic Circle at Kiruna, Sweden, researchers measured an ozone deficit very similar to the initial stages of ozone depletion in early September in the Antarctic. With so much extra chlorine in the stratosphere, repeated occurrences of such winters could cause sudden ozone reductions over the Arctic and perhaps over much of the Northern Hemisphere.

Although the unusual chemistry of polar stratospheric clouds has made the Antarctic ozone layer more vulnerable than the rest of the atmosphere, there is the particularly troubling possibility that similar chemical reactions could occur in warmer latitudes. Temperatures outside the polar regions are 20° to 30°C too warm for ice clouds to form, but droplets of sulfuric acid and water can support reactions involving the same chlorine reservoir compounds that deplete stratospheric ozone over the Antarctic and may help to explain part of the 3 percent ozone decrease observed over the Northern Hemisphere in the past two decades.

One prospect is that sulfurous particles emitted by a large volcanic eruption could team up with chlorine compounds to accelerate ozone destruction. Solomon and David J. Hofmann, of the University of Wyoming, describe a sharp drop in stratospheric ozone at mid-latitudes in 1982 after El Chichon erupted in Mexico, vaulting tons of volcanic debris into the upper atmosphere. At the time, the ozone drop was unexplained; atmospheric chemists still thought in terms of gases, not surfaces of particles. In light of the recent ozone studies, it seems likely that the sudden increase in the availability of surfaces provided by the volcanic debris allowed the industrially produced chlorine compounds to break down into chlorine atoms that could then destroy ozone, though more slowly than in the Antarctic.

Many factors other than industrial chemicals affect the concentration of stratospheric ozone. Ozone ebbs and flows along with the cycle of sunspots. This solar cycle affects ozone because during the height of sunspot activity ultraviolet radiation increases at wavelengths that can split apart an oxygen molecule to form a molecule of ozone, causing a change of a few (1 to 2) percent in ozone concentrations. The solar cycle was winding down between 1979 and 1986, but it is currently increasing. The upswing in sunspot activity will lead to ozone production that could partially cancel the chlorine-caused decline, but this will be temporary. The researchers warn against a sense of false security: After 1991 ozone could decrease again. They also suspect that ozone responds to a 26- or 27-month cycle of varying wind direction in which shifts in winds from the equatorial stratosphere change the flow of ozone to the poles.

Still other factors fuel concern for the global ozone layer. Rowland and colleagues report that the amount of water in the normally arid stratosphere could increase by 25 percent by the middle of the next century (because of water vapor produced with oxidation of increasing amounts of methane in the atmosphere) and contribute to increased cloud formation.

EFFECTS ON LIFE

The ozone layer is essential to life because it shields it from damaging ultraviolet radiation. Ironically, much less is known about the biological effects of increased ultraviolet radiation than about the chemical processes of ozone depletion in the atmosphere. Researchers are trying to learn how humans, vegetation, and aquatic ecosystems each may be affected by ozone depletion.

Scientists do know that direct exposure to ultraviolet radiation can damage the human immune system, cause cataracts, and increase the incidence of skin cancer. The EPA estimated in 1986 that the incidence of skin cancers would rise 2 percent for each 1 percent depletion of stratospheric ozone. (Today, mostly because of lifestyles that encourage skin exposure to strong sunlight, there are about 300,000 to 400,000 new cases of skin cancer each year in the United States.)

As part of the effort to understand the effects on vegetation and crops, researchers have tested more than 200 plant species, two thirds of which show sensitivity to increased ultraviolet exposure. Soybeans, one of civilization's staple food crops, is particularly susceptible to ozone damage, as are members of the bean and pea, squash and melon, and cabbage families. Plant responses to ultraviolet radiation include reduced leaf size, stunted growth, poor seed quality, and increased susceptibility to weeds, disease, and pests.

Scientists are also in the early stages of understanding how ultraviolet radiation might affect marine ecosystems and animals. Concern about these systems begins with phytoplankton, microscopic marine algae that form the base of the marine food web. Studies in the tropics have shown that significant amounts of ultraviolet radiation can kill them, while lesser amounts can slow photosynthesis and thus productivity. In Antarctica, this could affect krill, tiny crustaceans a notch up the food chain, and then fish, birds, and marine mammals including seals and whales. While water provides some protection from radiation, crude estimates indicate that ultraviolet radiation can penetrate to depths of 10 to 20 meters. Some phytoplankton are known to be tolerant of ultraviolet radiation, whereas others cannot tolerate any. A likely response will be for tolerant species to replace sensitive ones, though no one knows how this would affect the fish that eat them.

NATIONS JOINING TO PROTECT THE OZONE LAYER

The strong scientific consensus that CFCs deplete the ozone layer prompted nations to come together in unprecedented cooperation. The Montreal Protocol on Substances That Deplete the Ozone Layer, negotiated in September 1987, calls for a 50 percent reduction in CFC production from 1986 levels by 1999. Forty-nine nations—including Canada, the United States, Japan, and many nations in Europe, which together consume 80 percent of the chemicals controlled—have ratified the protocol.

An important factor in the discussion leading to the protocol was the recognition that, because the chlorine compounds are so stable, CFC molecules emitted today will exist to deplete ozone for a century or more. The average lifetime of CFC-11, for instance, is believed to be about 75 years and for CFC-12, 110 to 140 years. With a 100-year average lifetime, Rowland explains, 37 percent of the CFCs will still be in the stratosphere after 100 years, about 13 percent after 200 years, and about 4 percent after 300 years. Researchers agree that CFC concentrations will continue to increase for 10 to 20 years after we stop releasing them to the atmosphere because they will escape from existing reservoirs such as automobile air conditioners, and because of the lag between emission, arrival in the high stratosphere, and decomposition. Thus, if the nations that ratified the protocol comply with the terms established, average global ozone losses will still continue, but at a slower rate. These facts, and the growing body of scientific data on the threat to the ozone layer, are prompting nations to consider a 100 percent reduction in CFC production by the year 2000.

The protocol is a delicate balance between the most up-to-date scientific information, reliable industrial expertise, and committed political leadership, all supported by strong and informed public interest. The Montreal Protocol may prove to be a model for actions that span national boundaries and interests as the world addresses common environmental issues such as greenhouse warming and other forms of global change. It is perhaps the best illustration of the emerging role of scientific information and scientists in discussions about policies to manage global change. As Norway's former Prime Minister and chairperson of the World Commission on Environment and Development Gro Harlem Brundtland explains, "The scientist's chair is now firmly drawn up to the negotiating table, right next to that of the politician, the corporate manager, the lawyer, the economist, and the civic leader."

10

Vanishing Forests and Vanishing Species

Human activity has altered nearly every landscape on earth, but none so extensively as the expansive tracts of forest that have been cleared for agriculture, settlements, and logging. Much of today's farmland in North America was covered with forests only a few hundred years ago. Forest clearing for agricultural and industrial purposes had created the pastoral, managed landscapes of France, England, and Germany by the eighteenth and nineteenth centuries.

While forest clearing still continues in some parts, Europe and North America have more forests overall than they did a century ago. As people migrated from rural areas to cities in the transformation from an agrarian to an industrial society, the farms they abandoned often reverted to woods.

Recently, scientists and the public have focused their concern on deforestation in the tropics, particularly because tropical forests contain more than half of the world's plant and animal species and because tropical soils are notoriously unproductive once cleared of their forest cover. They are also concerned because as trees are cut and burned, or as they decompose, they release carbon back to the atmosphere, adding to the carbon released as humans burn fossil fuel to satisfy their energy needs.

Deforestation in the tropics has accelerated dramatically since World War II, but estimates of the area covered by the world's tropical forests and the rates of deforestation vary widely, mainly because countries use different survey procedures and definitions of "forest" and because satellite images of the entire globe are expensive and difficult to analyze. Simply put, we do not have reliable and up-to-date information on how much of the earth's surface is covered by forests and how fast it is being cut down.

The most extensive assessment of the state of tropical forests was a 3-year investigation in the late 1970s by the U.N. Food and Agriculture Organization and the U.N. Environment Programme. The study revealed that just three countries—Brazil, Indonesia, and Zaire—contain a major share of the world's tropical forests. The study focused on two basic forests types. The first, closed tropical rain forests, have a relatively tight canopy of mostly broad-leafed evergreen trees sustained by 100 centimeters (256 inches) or more of annual rainfall. The second, open tropical forests, have a canopy that is not continuous but covers more than 10 percent of the ground. When both types are considered, Brazil contains 26.5 percent of the world total, Zaire 9.2 percent, and Indonesia 6.1 percent. Peru, Angola, Bolivia, and India each have about 3 percent. The remainder is distributed among some 70 other tropical countries.

The study concluded that about 11 million hectares (about 27 million acres) of tropical forest is disappearing each year and that rates of deforestation vary greatly from country to country. In Zaire, for instance, deforestation rates were as low as 0.2 percent per year (which amounts to 1.8 million hectares deforested each year), but in the Côte d'Ivoire rates were as high as 7 percent (which amounts to 2.9 million hectares). Based on these rates, tropical rain forests would disappear altogether within 25 years from four countries in the Americas, three countries in Africa, and two countries in Asia. Within 50 years, tropical rain forests would vanish from 13 additional countries.

Philip Fearnside, of the Brazilian National Institute for Amazonian Research, analyzed satellite photographs to decipher deforestation trends in the Brazilian Amazon. His analysis indicates that deforestation in the Brazilian Amazon is highly concentrated in areas where colonization and immigration are intense, such as Rondonia, and that rates of deforestation increased exponentially between 1975 and 1980. Eneas Salati, a hydrologist and climatologist at the University of São Paulo in Brazil, and his colleagues calculate that if current trends continue, forests will be completely cleared from most of the Amazonian states by the year 2000.

WHY ARE THE FORESTS DISAPPEARING?

In developed countries, the competing interests of loggers, developers, naturalists, hikers, and hunters pose a challenge for forest management, but these conflicts pale in comparison with the challenges in developing countries. In poor countries, hundreds of millions of people depend on the resources from the forest for their very survival—land for their crops, fuel wood for cooking, and fodder for their animals. Governments, too, look to forests to provide much-needed foreign exchange as demand grows in the richer developed countries for exotic tropical hardwoods.

The reasons for deforestation in the tropics are many. Rapidly increasing populations need land on which to grow food. Land is concentrated in the hands of a few, leaving millions in search of unclaimed parcels. Roads and highways are built, making the forests accessible.

One of the agents of deforestation is shifting cultivation, a practice in which subsistence farmers clear and burn a plot of land in the forest, grow crops for one or a few years, and then allow the land to remain fallow for several years before repeating the cycle. This age-old method of subsistence agriculture recycles nutrients to the soil and maintains productivity of the land without fertilizers, provided the fallow period is long enough to regenerate the forest growth. But with increasing population and pressure on the land, in many places the fallow period is cut short. Eventually, the soil becomes unproductive, crops no longer flourish, and the trees do not grow back. The assessment by the U.N. Food and Agriculture Organization and the U.N. Environment Programme in the late 1970s estimated that shifting cultivation causes 70 percent of deforestation in Africa, 50 percent in Asia, and 35 percent in the Americas.

In densely populated countries, such as Ethiopia and India, with little or only seasonal rainfall and open woodlands, deforestation occurs as people gather firewood for fuel and fodder for animals. The U.N. Food and Agriculture Organization estimated in a 1983 report that, worldwide, three quarters of the 2 billion people who rely mostly on wood for fuel are cutting wood faster than it is growing back. With less wood available, women, the traditional fuel wood gatherers, have no option but to collect cow dung, which once would have fertilized the soil, for cooking fuel. Deprived of nutrients, the ability of the land to regenerate forests is compromised further.

In the Amazon, the opening of highways over the last 20 years promoted new and easy access to forested areas. The expanded access, combined with government incentives for development of the region, made it possible for the activities associated with rapid deforestation to take place—land speculation, cattle ranching, timber extraction, clearing for cash crops like rubber, cocoa, and sugar cane, and exploration for oil and minerals.

Government policies, too, have a large influence on how forests are used. Robert Repetto, an economist at the World Resources Institute, concluded in a 1988 report that tax and trade regimes, land tenure laws, agricultural resettlement programs, and administration of timber concessions with loggers are but a few of the policies that aggravate deforestation. He finds that these policies can contribute significantly to the wasting of forest resources. In the Brazilian Amazon, for instance, generous tax credits created over 12 million hectares of large cattle ranches, even though most of the ranches would have been unprofitable without these subsidies.

SOIL EROSION, RAIN MACHINES, AND CARBON

Why are forests such an important part of the earth system? On a local scale, trees protect the soil from rain and wind that would otherwise wash or blow it away. Despite the image of luxuriant growth in tropical forests, most of the soils that support them are remarkably unproductive. High temperatures and rainfall throughout the year encourage leaching of nutrients from the soil, so that few nutrients remain except for those held by the plants themselves.

Once forests are cleared for agriculture, grazing, or logging, there is no guarantee that the trees can grow back on the impoverished soil. Haiti, for example, deforested over the centuries, sports a landscape dominated by sparse grasses and bare mineral soil and bedrock. In Haiti and other areas where deforestation has been extensive, such as in the Atlantic coast of Brazil and the mountains of southern China, not only may the soil be damaged but there are few adult trees to provide seeds for new forests.

Soil erosion is a natural process. Without it, deltas would not form as soil erodes from the land and travels as sediment through streams and rivers. But the soil exposed in a deforested site greatly accelerates this natural process, so much so that some dams have filled with sediment far more rapidly than expected. Salati explains that, although there are very few studies of soil erosion and river sediment loads in tropical areas, the few existing data do show that erosion losses can be 100 times greater in soils changed to agricultural use when compared to similar soil covered with forest.

On a regional scale, forest ecosystems recycle the rainwater back to the atmosphere through evaporation from the soil and leaf surfaces and through transpiration from plants, a process so efficient that ecologists refer to tropical forests as "rain machines." The rainfall patterns in the Amazon basin, and probably other regions such as the Brazilian central plateau, depend on the existence of the forests. Salati and colleagues, studying the effects of deforestation on the recycling of rainfall in the central Amazon, point out that in a healthy stand of rain forest, about half of the rainfall is evaporated from the surface of soils and leaves or transpired by plants and the other half runs into streams and rivers, eventually returning to the ocean. With deforestation, this vigorous recycling of water will weaken and could lead to lower rainfall in the region.

The rapid runoff of rainfall from deforested areas has led some scientists to link the 1988 flood in Bangladesh to deforestation in the Himalayas. In the worst flood in Bangladesh's history, two thirds of the country was under water for several days, and 25 million of the country's 110 million people were left homeless. The researchers hypothesize that deforestation has made the upland watersheds of the Himalayas less able to store water and moderate the flow of water into the streams, in turn accelerating the flow of water into the Brahmaputra River and exacerbating flooding as the river flows into the Bay of Bengal.

Forests assume an essential role on the global scale. The forest cover absorbs energy that would reflect back to the atmosphere if the soil were bare. Plants take up carbon from the atmosphere as they grow, and release carbon back to the atmosphere when they are burned or die and decompose. On balance, if the amount of forest cover were to remain constant, the uptake would equal the release of carbon over the long term. But as more trees are cut than are planted worldwide, more carbon is released to the atmosphere than is stored, adding to the carbon dioxide being released from the burning of fossil fuels for energy. Salati, with colleagues Reynaldo Luiz Victoria, also of the University of São Paulo, Luiz Antonio Martinelli, of the Centro de Energia Nuclear na Agricultura of Brazil, and Jeffrey Richey, of the University of Washington, find, based on a large range of estimates about rates of deforestation and how much biomass the forests contain, that annual emissions of carbon dioxide from deforestation in the Amazon alone account for 4 to 25 percent of carbon dioxide emissions to the atmosphere worldwide. Other estimates indicate that deforestation could account for about 20 percent of the buildup of carbon dioxide in the atmosphere.

The difficult task of estimating how much carbon is entering the atmosphere because of deforestation is complicated further because plants themselves respond positively to increased concentrations of atmospheric carbon dioxide. Like sunlight, water, and nutrients, carbon dioxide is required for plant growth. With an increase in atmospheric carbon dioxide, growth can proceed more rapidly, which in turn would theoretically increase the amount of carbon dioxide that trees could remove from the atmosphere. Researchers have observed that in greenhouses growth rates do increase over the short term when plants are fumigated with carbon dioxide. But there have been very few long-term experiments on mature trees in their natural settings. Quite simply, the question of whether increased concentrations of atmospheric carbon dioxide will stimulate plant growth enough to offset some of the carbon being released to the atmosphere from deforestation is an open one.

Another open question is how global warming itself could alter the distribution of forests and consequently change the amount of carbon taken up from the atmosphere. Climate modelers generally agree that temperature increases from human-induced global warming would be greatest in the high latitudes. The tropics would experience only modest temperature increases. So those forests in the high latitudes of Canada, Alaska, the Soviet Union, and Scandinavia might expand northward into areas previously covered by tundra vegetation. Areas now covered with grasses would store more carbon in their biomass if they became forests. But these possibilities are speculative and require a great deal of scientific investigation.

Because of the important role of forests in determining the amount of carbon in the atmosphere, some researchers argue that reforestation could help counteract the buildup of atmospheric carbon dioxide from human activities. Roger A. Sedjo, a senior fellow at Resources for the Future, and colleagues calculate that new forest plantations covering an area of approximately 465 million hectares would be required to remove 2.9

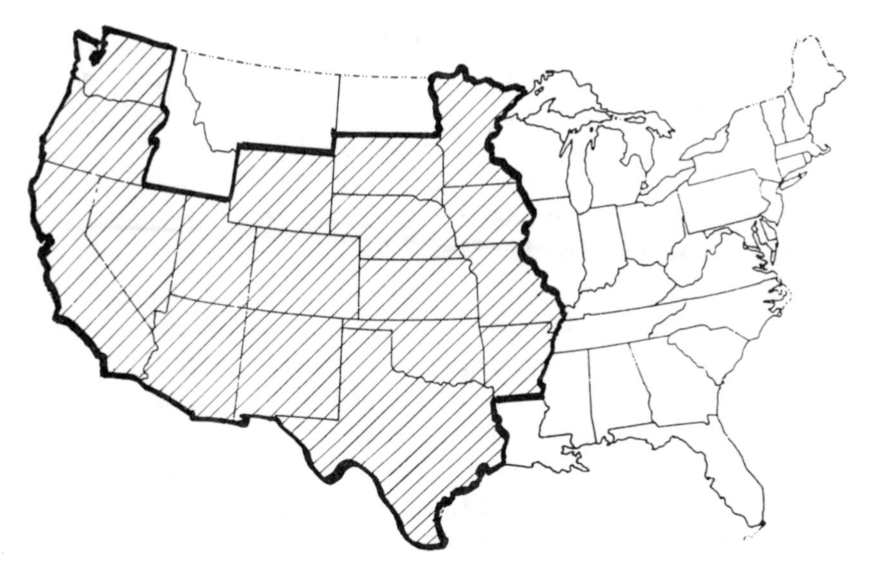

An area approximately equal to the 465 million hectares of newly planted forest that would remove the 2.9 billion tons of carbon added annually to the atmosphere. (Reprinted, by permission, from Norman J. Rosenberg et al., eds. 1988. *Greenhouse Warming: Abatement and Adaptation*, Fig. 8-1, p. 114. Copyright © 1988, Resources for the Future.)

billion tons of carbon, the amount of carbon accumulating each year in the atmosphere. It is unlikely that trees will be planted on this scale, but reforestation projects could make modest dents in the accumulation of carbon dioxide and, at the same time, provide other benefits for water quality and for species that live in the forests.

VANISHING SPECIES

It is nearly impossible for a biologist to be present at the precise moment when the last individual of a species languishes in a tropical forest. Nevertheless, with the clearing of tropical rain forests for agriculture, grazing, and other uses necessary to support the world's population at a satisfactory standard of living, we know that habitats are destroyed and many species of plants and animals are irreversibly lost before scientists can even identify them.

Human civilization has always been dependent on the natural diversity of living organisms. In the course of history, people have exploited about 7000 species for food. But the species that we have used to support the human population are only a small fraction of the total number of species on earth. Some 1.4 million species have been identified and given scientific names, but scientists estimate that millions more have not yet been identified, particularly in the species-rich tropics. Until the 1980s, estimates of the total number of species on earth ranged between about 3 million and 10 million species. Terry L. Erwin, of the Smithsonian's National Museum of Natural History, later discovered huge numbers of insect, particularly beetle, and other arthropod species in the forest canopies of moist tropical forests in Brazil and Peru, causing him to revise his estimate of the worldwide number of insect species alone to 30 million.

Scientists and the public worry that, with deforestation and the loss of natural habitat, many of these species will be gone before they are even known to exist. Trends in the moist forests of the tropics, which cover only 7 percent of the earth's land surface but contain over half of the world's species, are especially of concern. These rain forests are incredibly diverse and rich in species, particularly species of insects and flowering plants, though ironically the species that live in rain forests are the least well identified. Thomas Lovejoy, of the Smithsonian Institution, estimates that between 15 and 20 percent of all species will become extinct by the year 2000 because of the destruction of tropical forests.

Ecologists have identified "hot spots" around the world where habitats rich in species are in imminent danger. Several areas stand out, according to Norman Myers, an environmental consultant with extensive experience in the tropics, including the Choco forest of Colombia, the Napo center of diversity in Peruvian Amazonia and other centers around the fringe of the Amazon basin, the Tai Forest of Côte d'Ivoire, the montane forests of East Africa, the relict wet forest of Sri Lanka, the monsoon forests of the Himalayan foothills, northwestern Borneo, certain lowlands of the Philippines, and several islands of the South Pacific.

Why does it matter if a lone, unidentified species should go extinct? After all, extinctions have occurred in the past. The history of life is punctuated by five massive extinction episodes, the last of which is most noted because it marked the end of the age of the dinosaurs and the beginning of the age of mammals. Such extinctions changed the course of evolutionary history, and the extinctions caused by humans will be no exception. Edward O. Wilson, a biologist at Harvard University, estimates that the rate of species loss from deforestation is about 10,000 times greater than the naturally occurring background extinction rate that existed prior to the appearance of human beings.

From a human point of view, the loss of a species is the loss of a potentially valuable contribution to humanity. At least 75,000 plant species have edible parts, for instance. Some of these species are superior to those widely used today. The tropics have provided the world with most of its edible species by far. Few people realize, as Mark Plotkin, of Conservation International, points out, that a typical American breakfast of cornflakes, bananas, sugar, coffee, orange juice, hot chocolate, and hash brown potatoes is based entirely on plant species that originated in the tropics. Many currently underexploited tropical species could become familiar sights in U.S. supermarkets. Other tropical species provide important ingredients for pharmaceuticals, fibers, and petroleum substitutes, like the rosy periwinkle of Madagascar, which provides the basis for an extremely effective medication for Hodgkin's disease and acute lymphocytic leukemia, and the babassu palm of the Amazon basin, which produces oil. No one knows how many other species have the potential to provide similar benefits to society.

The diversity of species and populations also constitutes the world's available gene pool, which in the words of Wilson "is one of our planet's most important and irreplaceable resources." Every microorganism, animal, and plant contains from 1 million to 10 billion bits of information in its genetic code, the product of thousands or millions of years of evolution. Scientists

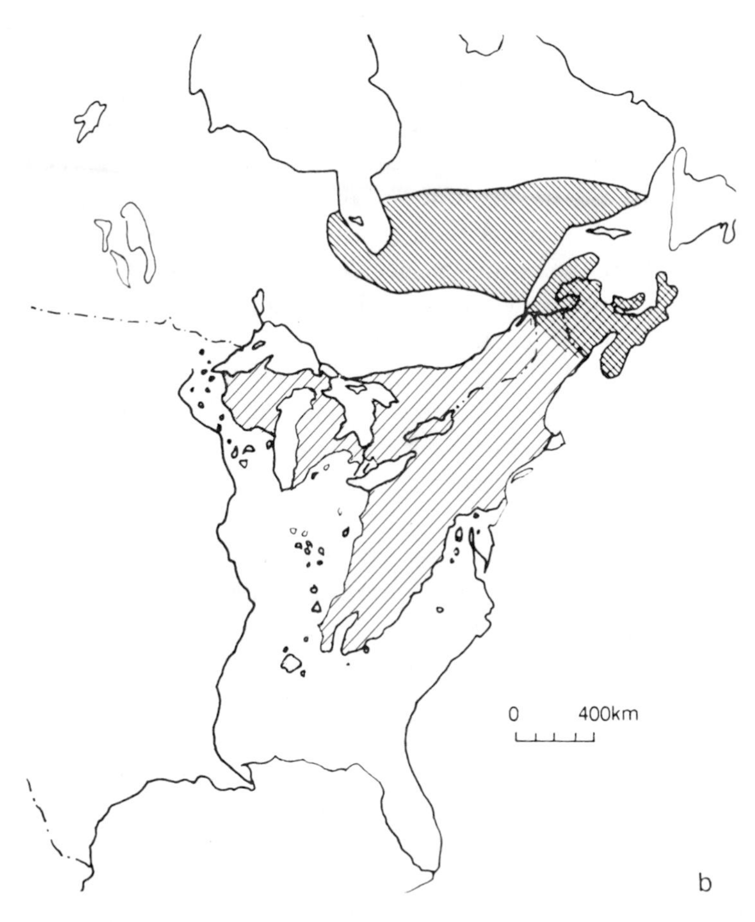

of favorable environment. Significant advance to the north is unlikely, as rates in the past were about 25 kilometers per 100 years, and the most rapid rate known from the fossil record (for spruce) is only 200 kilometers per 100 years. As a consequence, much of the potential range will remain unoccupied. (Reprinted from *Toward an Understanding of Global Change*. National Academy Press, Washington, D.C., 1988.)

take advantage of this genetic richness to improve those species that are used by humans. Commercial species are continuously crossbred with their wild relatives to improve yield, nutritional quality, responsiveness to different soils and climate, and resistance to pests and diseases. Because many of the world's important crop species originated in the tropics, their relatives that provide the means to maintain and improve our crops can only be found there. A wild relative of the domestic tomato—discovered in 1962 quite by accident while Hugh Iltis, director of the University of Wisconsin Herbarium, was waiting to pass a landslide on an excursion in the Peruvian Andes—after almost two decades of research increased the sugar content of the tomato and raised its commercial value by millions of dollars. Other species provide singular genetic material for use in medicines. The birth control pill, for example, has its origin in the Mexican yam.

Apart from the reasons for maintaining tropical species solely for the purposes of human exploitation, many take the more philosophical view that humans are stewards of the planet and have responsibility for maintaining its diversity. Instead we are eradicating in the blink of an evolutionary eye species that have evolved over the course of millions of years.

Even as deforestation threatens biodiversity, the predicted changes in climate portend more stress on the world's plant and animal species. When temperature and precipitation patterns change, so does the geographic range over which a particular species can survive. Robert Peters, an ecologist with the World Wildlife Fund, explains that even very small changes of less than 1°C have caused substantial changes in range during this century. The white admiral butterfly and the comma butterfly, for example, greatly expanded their ranges in the British Isles during the past century as the climate warmed about 0.5°C, and the birch responded rapidly to warming during the first half of this century by expanding its range north into the Swedish tundra.

Unlike animals, plants must rely on mechanisms that disperse their seeds—mainly animals and winds—to adapt to changing conditions. Margaret Davis and Catherine Zabinski, both of the University of Minnesota, forecast the future response of some hardwood species—eastern hemlock, yellow birch, beech, and sugar maple—to the expected changes in climate. On the basis of climatic conditions projected by two of the general circulation models for an effective doubling of carbon dioxide, the researchers predict that the species would need to migrate northward by 500 to 1000 kilometers (310 to 620 miles) to remain in their suitable climate. Beech, for instance, which now grows east of the Mississippi River from southern Canada to Florida, would have to shift into New England and southeastern Canada as far north as Hudson Bay.

The key question is: How quickly can a species migrate as its suitable habitat changes? Some 10,000 to 12,000 years ago as the last glaciers retreated and temperatures rose 3° to 5°C (5.4° to 9°F), species of forest trees migrated at rates of 25 to 40 kilometers per century, with the fastest migration by spruce into northwestern Canada at 200 kilometers per century. Analyses of fossil pollen grains indicate that beech forests moved about 20 kilometers per century, far less than Davis and Zabinski indicate would be necessary to keep pace with climate changes projected for the coming century.

The rates of climate change predicted for the next century are possibly 40 times faster than at the end of the last ice age. No one knows whether plant species would be able to keep up. For certain, each individual species would respond in its own way. Some may go extinct, and others may survive in reduced numbers or with changes in their geographic distributions. As various species respond, they would be likely to form communities of plant species that are different from those that we have today.

Even species that could keep pace with climate change would encounter barriers as they migrated. Cities, oceans, or mountain ranges could get in the way, especially in the case of species isolated in reserves designed for the very purpose of conserving them. As Peters explains, conservationists have heretofore managed an unchanging world. "Now we know this is not so," he notes. "The world changes and our conservation plans must change with it."

PROTECTING THE FOREST RESOURCES

While policymakers and the world's leaders struggle to address the root causes of deforestation, successful projects around the world demonstrate that it is possible to reconcile economic development with preservation of tropical forests.

The project started in 1984 to protect the Khao Yai National Park in northeastern Thailand and simultaneously to provide benefits to villagers living nearby is one such example. Wild elephants were a central theme in the conservation plans. Not only have the elephants been protected, they also have attracted tourism, which benefits the villagers. In 1985 a project on "rural development for conservation" was begun in the nearby village of Ban Sap Tai to offer agricultural assistance, credit, health care, and education. The improvements in economic conditions have reduced the villagers' dependence on the park and their incursions to cut trees, plant maize, and poach plants and animals.

Similar projects have been implemented that preserve forests while allowing the land to be used for agriculture, forestry, or watershed management. Such projects illustrate that the technology and knowledge exist to preserve the world's tropical forests. Judith Gradwohl and Russell Greenberg, both of the Smithsonian Institution, analyzed case studies of such successful projects. They find that the projects are most likely to succeed when local people participate in the planning and implementation and when the plan itself is both environmentally and economically sound. For example, forest reserves can offer an economic return through sustainable agriculture and forestry and through gathering of nonwood forest products such as latex, fibers, fruits, nuts, and a host of other miscellaneous products. Gradwohl and Greenberg explain that most environmentally oriented projects will not produce as much immediate economic gain as ecologically disastrous exploitation such as cattle ranching and uncontrolled timbering. But "if the goal changes to include both profits and well-being of forests, then ecologically based schemes compare very favorably."

11

Lakes, Forests, and Acid Deposition

Even though the British scientist Angus Smith coined the term "acid rain" over a century ago, only in the last few decades have scientists recognized that widespread acidity in precipitation causes damage far from its source. Over large stretches of the world, acid deposition has damaged life in lakes and streams and corroded building materials and accelerated the aging of structures. In addition, it has become a key suspect in the declining health of some species of forest trees in North America and Europe.

Acid deposition results when pollutants, particularly oxides of nitrogen and sulfur, are emitted from smokestacks, smelters, and automobile exhausts into the atmosphere. These oxides are converted, through a series of chemical reactions with other substances in the atmosphere, to acids that fall back to the earth's surface dissolved in rain, snow, or fog, or as gases or dry particles.

The political tensions surrounding acid deposition arise largely because the effects of pollutants produced in one region can be felt in another. Lakes in far-upstate New York are acidified, in part, by emissions from the smokestacks of midwestern power plants. Acids that rain into Scandinavia originate in central Europe or the United Kingdom, and about 50 percent of the acid deposition falling in eastern Canada comes from the United States.

Acid deposition has been detected recently in other industrialized areas, including western North America, China, Japan, the Soviet Union, and South America. In some areas of Africa that are not heavily industrialized, high levels of nitric oxide and other gases implicated in acid deposition have been measured. Fires set by farmers to clear the forests and savannah are a possible source.

The main cause of acid deposition in the industrialized world is the sulfur oxide emitted to the atmosphere when coal is burned as fuel or when high-sulfide ores are used in smelters. The amount of sulfur in coal varies from deposit to deposit. The higher the sulfur content, the greater is the contribution to acid deposition once the coal is used as fuel. Not all coal is high in sulfur. Coal from the midwestern United States is high in sulfur—about 3 percent by weight. Coals from Appalachia contain from 1 to 3 percent sulfur. Western coal has relatively little sulfur, less than 1 percent. China, which is embarking on ambitious plans for industrial development, will build factories and power plants drawing on that country's large reserves of high sulfur coal. Coal that contains high concentrations of sulfur is also burned elsewhere in the world, such as in some Eastern European countries.

Nitrogen oxides are the second major source of acidifying compounds. Nitrogen oxides are emitted as a by-product when fossil fuels like gasoline, oil, and natural gas are burned. The amount emitted depends on a variety of factors, particularly the temperature of combustion. A large fraction of the nitrogen oxides responsible for acid deposition is emitted from automobiles and other vehicles. Stationary sources such as power plants also contribute significant amounts of nitrogen oxides to the atmosphere.

Scientists and the public in Europe became increasingly

aware in the 1960s and 1970s that the amount of dissolved acids in precipitation depends on the direction, timing, and speed of air flowing over Europe and England. In North America and the rest of the world, concern took longer to set in. Throughout the early 1970s, research in North America on air pollution was motivated more by interest in the potential effects of atmospheric pollutants on human health than by concerns over the effects on ecosystems in water or on the land. Gradually it became clear that changes in the chemical composition of precipitation were having—or had the potential to have—significant effects on ecosystems. Since then, scientists have conducted millions of measurements and produced thousands of publications to understand the causes and consequences of acid deposition.

THE ACIDIFICATION PROCESS

Nearly all of the acidification causing environmental damage comes from sulfur dioxide (SO_2) and nitrogen oxides (NO_x) released as gases when fossil fuels, particularly coal, are burned. In the atmosphere, these oxides are transformed through a series of chemical reactions into sulfuric and nitric acids (H_2SO_4 and HNO_3) and then into sulfate and nitrate. The rates of transformation are governed by environmental conditions such as sunlight, temperature, humidity, clouds, and the presence of various other chemicals. The acids dissolve in water droplets and eventually rain back to the earth.

The processes that govern how quickly the acids or acidifying substances are deposited on the earth, and how far they travel in the atmosphere, are complex and depend on meteorological events, the characteristics of the earth's surface, and the particular form of the pollutant. Typically, the pollutants remain in the atmosphere for periods ranging from hours to weeks, in which time they may travel distances from a few to over a thousand kilometers.

Although concern about acid deposition has focused on wet forms (hence the term acid rain), the acids can also be deposited "dry." Less is known about this process, mainly because it is very difficult to measure, but evidence indicates that when the dry material comes in contact with moisture, it causes the same environmental effects as acids in rain, snow, or fog. In addition, sulfate in the atmosphere that takes the form of fine particles scatters light and reduces visibility.

The degree of acidification is measured by the pH value on a scale of 0 to 14, which depends on the concentration of positively charged hydrogen ions in solution. A neutral solution such as distilled water has a pH of 7. Acidic solutions have higher concentrations of hydrogen ions, indicated by lower pH values. The lower the pH value, the higher the acidity.

Most natural waters, including rain and snow, are somewhat acidic because of naturally occurring chemicals. In remote areas untouched by anthropogenic sulfur and nitrogen, rain with pH values of 5.2 to 5.4 is common, and values of less than 5 have been recorded in extreme cases. But the organic acids and small amounts of naturally occurring nitric and sulfuric acids responsible for this acidity do not have the same consequences for life in lakes and streams and on the land as do the mineral acids from acid deposition.

Scientists have determined that in soils receiving doses of acid deposition, the excess hydrogen ions displace other elements, including nutrients such as potassium, magnesium, and calcium, and retard plant growth. High concentrations of hydrogen also release aluminum, a metal naturally occurring in soil. Once released, the aluminum can be toxic to plants by interfering with the ability of roots to absorb water and nutrients. Acidity in water can cause lakes and waterways to lose their abilities to support species that thrived under less-acid conditions.

How ecological systems respond to the addition of acidic material depends on the rate at which it is deposited and on a region's geologic composition. Some of the rainwater entering a lake or stream falls directly on the water's surface, but most falls on the soil surrounding the water body and travels over the surface and through the soil. Minerals in the soil can react with the hydrogen ion to neutralize the acidity of the water while it is en route to the lake or stream. This buffering capability of the soil can slow down or prevent the acidification of lakes and streams, but only where the particular type of soil contains the right types of minerals. Sandy soils or soils that are thin and highly weathered generally are already acidic, and so they have only limited abilities to neutralize the effects of acids deposited from the atmosphere. Soils capable of retaining large amounts of sulfate and nitrate, or soils rich in elements such as calcium that can neutralize the acid, can forestall the input of acidic water to lakes and streams, at least until the buffering capacity is used up. In the lake or stream itself, acid-neutralizing chemical reactions and minerals in the water and sediment can further counteract the acidity, but with heavy inputs and with time some of this capability can also be exhausted.

Because of the makeup of the soils, extensive regions of North America and northern Europe are sensitive to acid deposition. Researchers are finding that more and larger areas are more vulnerable than they believed 10 years ago. In the United States, these areas occur in the Northeast, as well as in Minnesota, Wisconsin, upper Michigan, parts of the Southeast, and many mountainous areas of the West. Surveys reveal that in Canada, half of the 700,000 lakes in the six eastern provinces are extremely sensitive to acid deposition. So are large stretches of Canada's western and northwestern provinces, as well as much of northern Europe and vast expanses of Asia, Africa, and South America.

LAKES

As researchers try to understand how acidification affects ecosystems, they are hindered by the lack of comprehensive biological surveys of large acid-sensitive areas, including North America. As David W. Schindler, an ecologist at the University of Alberta in Edmonton, Canada, explains, "As a result of never knowing what we had, we cannot know what we have lost."

Studies of the effects of acidification have been based largely on observations of lakes where little is known of their state before air pollution became a significant influence on them. As a result, in many of the lakes known to have been affected, only obvious changes such as the disappearance of large game fish have been recorded. The surveys indicate that the sensitivity of lakes to acid deposition varies from place to place. Even neighboring lakes can respond differently.

Much of the understanding about the effects of acid deposition on lake ecosystems has been obtained from studies in laboratories and artificial acidification of experimental lakes. Lakes in or near populated areas may be affected by numerous chemical pollutants from a variety of sources including recreational uses, agriculture, and nearby mining. Consequently, there are few pristine places where scientists can study the effects of acidity in the absence of these other influences. Recognizing this need, the Canadian government designated a portion of northwestern Ontario the Experimental Lakes Area. Scientists control 46 lakes and their watersheds, conducting experiments to help understand what happens when the chemistry of lakes changes. One of these lakes, "Lake 223," has provided especially valuable insights.

Lake 223 lies in a region with thin, sandy soils covered by pristine forests of jack pine and black spruce. It receives little acid deposition, and the surrounding watershed has very poor acid-neutralizing capacity. After 2 years of background study, researchers began in 1976 to acidify the lake experimentally, adding sulfuric acid incrementally until by 1983 they had lowered the pH of the lake to 5.0 from its original value of 6.8. In 1985, Schindler and his colleagues reported that the species living in the lake suffered the effects of acidification earlier than expected and that, even when the pH was relatively high, changes were extensive. The overall biological productivity and availability of nutrients were essentially unchanged, but a handful of species that were food for the healthy population of trout in the lake were eliminated at pH values as high as 5.8. When the study began, the lake supported a community of about 220 species. By the time the pH reached 5.0, fewer than 150 remained. (Even this number is misleading, Schindler says, because almost half of that number were acid-resistant organisms that had come in to replace ones that had been forced out by the acidity.) After 8 years, the trout—the top of the food chain in the lake—were no longer reproducing, and their mortality rates had gone up considerably.

Scientists believe that the number of species in a lake declines continuously with increasing acidity below pH values of 6.5 to 7.0 and that many species that are foraged by fish higher in the food chain are lost at pH values near 6.0. This disruption in the food chain means that large predatory fish can starve long before the direct toxic effects of acidification are evident.

"We are not losing hundreds of thousands of species, we are losing hundreds," Schindler says. "But in terms of the fraction of species that make ecosystems function, the relative magnitude of biological impoverishment in acidifying softwater lakes is probably just as large as in the tropics."

These findings hold true when researchers estimate the likely degree of biological impoverishment of lakes in the northeastern and midwestern United States. Such estimates are possible because, while there are too few data to indicate the number of species lost, scientists find they can use existing chemical data to predict the proportion of species lost as a lake moves from any given pH to a lower one. Schindler and colleagues predict that more than half of the lakes from the area's most heavily acidified regions have lost 40 or 50 percent of species such as molluscs and insects, losses just as important to the ecosystem (though not as visible) as losses in gamefish. Thorough case studies of streams, while few, indicate that the life they support may be even more sensitive to acidification than that in lakes.

While some lakes in Ontario provide an example for what happens under extreme acidification, others make it possible to study what happens when acid precipitation lessens. In 1972, emissions from the Coniston smelter, which contributed massive amounts of sulfur dioxide, were reduced by over 60 percent, and the three smokestacks less than 200 meters tall were replaced by one 381 meters tall. The combined effect of these changes was to give the life in the lakes near Sudbury a chance to recover.

(Overall, Canada has reduced its emissions by 50 percent, and in Ontario by nearly 60 percent, of the 1980 emission level.) Once additions of the acidic materials tapered off, alkalinity and pH increased rapidly, but the biology of the lakes was less resilient. While some species soon recovered, resuming reproduction when pH levels rose to between 5.4 and 5.6, some, such as trout, did not. Nor did all species eliminated from the lakes return; stocking of fish could be required to return the lakes to states close to their original ones.

FORESTS

Damage to trees from droughts, hurricanes, insects, and disease is as old as the forests themselves. The transport of pollutants from their industrial sources to stands of forests hundreds of kilometers away subjects these ecosystems to yet another form of stress.

In the late 1970s, extensive discoloration of needles and declining rates of growth in forests in northeastern Bavaria in the Federal Republic of Germany suggested to scientists that acid deposition, and the ground-level ozone produced by the chemical reactions involved in acid deposition, play a major role in what has come to be known as "forest dieback" or, in German, *waldersterben* (forest death). By the early 1980s, 20 to 25 percent of European forests were classified as moderately or severely damaged. In the late 1980s, the pace of deterioration decreased. The causes of the damage are not entirely clear. In the United States, long-distance transport of air pollutants is suspected of contributing to a decline in forest health in California's San Bernardino National Forest, in the pine forests of the Southeast, and in the high-elevation forests of the Appalachians.

All trees experience stress in the form of diseases, insects, extremes in the weather, and competition with other trees for light, nutrients, and water. When the accumulated stress becomes too severe, trees become more vulnerable to opportunistic pests or injury from extremes in the weather. Eventually, the growth rate

A stand of Douglas fir on Whiteface Mountain in the Adirondack Mountains, New York, suffers damage from a combination of factors, including disease, harsh winters, and acid precipitation. (Photo taken in 1987 by Y. Yee, U.S. Forest Service.)

for the tree declines, and, if the stress is severe enough, the tree dies.

At high enough concentrations, air pollutants such as ground-level ozone and sulfur dioxide can damage a tree's foliage when they come in direct contact with it. Indirectly, acid deposition could deplete the soil of nutrients essential for the tree's growth, elevate the levels of toxic metals such as aluminum, and alter the normal functioning and growth processes. Because clouds can contain high concentrations of ozone and acids, and because soils at high elevations are relatively thin, the effects on forests are most pronounced at high elevations, where clouds and fog come in contact with the trees.

It is difficult to distinguish between damage to trees from natural causes and from air pollutants. Recent analyses, notably by the U.S. government's National Acid Precipitation Assessment Program, have considered forest damage in North America and attempted to unravel the causes. Red spruce that grows at high elevations of the northern Appalachians in the eastern United States has drawn special attention because analyses of tree rings reveal that in the past quarter century, the growth rate of the trees has declined substantially and over half of the mature red spruce trees in the high elevations of the Adirondack and Green mountains have died. In New York, Vermont, and New Hampshire, stands of red spruce began to decline between the late 1950s and mid-1960s. There is much stress from natural factors at high elevations, but scientists have now focused their studies on effects of acidic cloud water and ground-level ozone, which may compound the other stresses to the point where red spruce is unable to survive. On the basis of current scientific understanding, an interaction between injury from air pollutants and mites seems a likely cause of the spruce decline in the mountain forests of the Northeast.

Recent research has led to new theories about the decline of Norway spruce in central Europe. A team of forest scientists led by Ernst-Detlef Schulze, of West Germany's University of Bayreuth, conducted a wide-ranging search for causes of the dieback in the Bavarian spruce forest but found no one agent responsible. His studies suggest that atmospheric deposition—nitrogen compounds in particular—gradually creates nutrient imbalances in the trees and soils. Eventually, the trees become deficient in important nutrients such as magnesium and more vulnerable to opportunistic blights and to extreme weather such as drought. Schulze hypothesizes that the visible symptoms of forest decline apparent since the late 1970s reflect this constellation of conditions.

Changes in climate as a result of the greenhouse effect would add to the stresses that air pollutants place on forests. Over short intervals such as seasons, of course, weather is a governing factor. But longer-term changes in temperature and precipitation would not only influence where and how rapidly acidic material is deposited, but would also change the environmental conditions for the trees.

BUILDINGS AND MONUMENTS

From an economic standpoint, corrosion of building materials is one of the most serious effects of acid deposition. According to the EPA, estimates of the annual costs of repairing or replacing structures damaged by acid deposition exceed $5 billion.

Marble and limestone, which is the second most commonly used building material in the United States, are particularly susceptible to acid deposition. The acids attack the calcium carbonate, the principal constituent of these materials. Limestone monuments like the Acropolis in Athens and the Jefferson Memorial in Washington, D.C., show signs of damage. Emissions from Mexico's refineries are corroding Mayan artifacts. In southwestern Colorado, air pollutants from several power plants and a nearby smelter are suspected in the quickening deterioration of ancient sandstone cities of the Ansazi Indians.

Of course, factors other than acid deposition contribute to aging and deterioration of materials, including sunlight, wind, and water. But evidence from buildings in areas receiving high levels of acid deposition indicates that the process is being accelerated. In addition to stone, acidic pollutants are also implicated in widespread damage to paint, wood, fabrics, masonry, concrete, and metals, though less is known about these weathering processes. The evidence is beginning to mount that rusted steel in bridges and corroded buildings are joining the list of the costs to society from acid deposition.

CONTROLLING ACID DEPOSITION

While researchers have been exploring the causes and consequences of acid deposition, engineers have been designing methods to control those emissions responsible for the acidity, and politicians have been debating the merits of implementing laws that would put these controls into practice.

Progress has been made over the last 15 years. In the northeastern United States, the sulfate content of rain and the concentration of sulfur compounds in the air have decreased, reflecting pollution control measures mandated by the 1970 Clean Air Act and efforts of individual states to limit emissions. Regulations for controlling emissions from automobiles have contributed substantially to the decline in nitrogen oxide emissions since their peak in 1978. But there is little doubt that emissions will have to be reduced much further to reduce the threats of acid deposition.

Implementing technologies to control emissions or setting restrictions on the sulfur content of coal raise a host of difficult and sensitive questions. On the scientific side: How much should emissions be reduced to protect the environment from the effects of acid deposition? In what regions should emissions be reduced to protect sensitive areas far from the source? Are technologies available to control the emissions? On the political side: Would different limitations on emissions for different regions be politically acceptable? Should controls be retrofitted to the older plants that predate the Clean Air Act, even though the remaining economic life of these plants is relatively short?

Should these older plants be decommissioned? Who should bear the costs?

Scientists have constructed computer models to help provide answers to some of these questions. On the basis of emission levels and forecasts of weather and atmospheric chemistry, the models predict the amount of acid deposition that would occur for hypothetical emission levels from different regions. Of course, the models can only reflect the state of the scientific understanding about acidification processes. Uncertainties such as how long the soil can buffer the effects of acid deposition, how long it would take for lakes to recover from the effects of acidity, and the relation between forest health and acid deposition need to be resolved before the models can provide more precise answers. Nevertheless, the models can help policymakers evaluate and balance issues related to strategies for controlling acid deposition. Where and by how much should emissions be reduced? What technologies should be used? Should laws be implemented now or should action be delayed until the remaining scientific uncertainties are resolved?

Reducing emissions from coal-burning power plants, which generate more than one half of the nation's electricity, is the key to controlling deposition of sulfuric acid. Techniques to reduce emissions are available for the 410 generating stations built between 1955 and 1975, when little was done to control emissions. Effective measures include switching to coal with lower sulfur content, "cleaning" the coal by removing the sulfur before combustion, and spraying the exhaust with wet limestone to neutralize the acid (a process known as flue-gas desulfurization). These methods have trade-offs. Fuel switching may displace coal miners; coal cleaning techniques typically remove only 10 to 30 percent of the sulfur; and conventional flue-gas desulfurization is expensive and reduces the efficiency of the plant. Moreover, none of these techniques do anything to reduce emissions of nitrogen oxides, also an important contributor to the buildup of greenhouse gases in the atmosphere.

As these older plants age, they become candidates for replacement or refurbishment and provide opportunities for incorporating one of the "clean coal technologies" that reduce emissions while maintaining, or even increasing, the efficiency of the plant. For instance, in the system known as atmospheric fluidized-bed combustion, combustion takes place at a lower temperature, which reduces the formation of nitrogen oxides. Meanwhile, limestone is mixed with coal in the combustion process and efficiently captures sulfur dioxide. Such techniques are improving and eventually could yield very large reductions in the emissions responsible for acid deposition.

AFTERWORD

Global Change and Our Common Future

by GRO HARLEM BRUNDTLAND

We live in an historic transitional period of burgeoning awareness of the conflict between human activities and environmental constraints. The world is finite, but it will have to provide food and energy to meet the needs of a doubled world population some time in the coming century. Its natural resources, already overtaxed in many areas, will have to sustain a world economy that may be 5 to 10 times larger than the present one. This cannot be done if humans continue to pursue current patterns.

As people continue their endless quest for new materials, new energy forms, and new processes, the constraints imposed by depletion of natural resources and the pollution caused by human activity have brought society to a crossroads. Abundance coexists with extreme need, waste overshadows want,

---

The Afterword is adapted from the keynote address and Franklin Lecture presented by Mme. Brundtland at the Forum on Global Change and Our Common Future, Washington, D.C., May 2, 1989. The views do not necessarily reflect those of the National Academy of Sciences.

and our very existence may be in danger owing to mismanagement and overexploitation of the environment. In spite of all the technological and scientific triumphs of the present century, there have never been so many poor, illiterate, or unemployed people in the world, and their numbers are growing. As they struggle to survive, they have little choice but to pursue activities that may undermine the environment, the natural resource base on which they depend, and the conditions that sustain life itself.

As the descriptions in this book make all too clear, we face a grim catalog of environmental deterioration. The very real possibility that our actions are depleting the earth's genetic resources, changing the climate and the composition of the atmosphere, and upsetting the chemical balance of our lakes and waterways proves that if we all do as we please in the short run, we will all lose in the long run. We need to develop a more global mentality as we chart our collective future, and we need not only firm political and institutional leadership but also sound scientific advice. Indeed, the role of men and women of science in shaping our future will become more central as the challenging dynamics of global change gradually become clear.

The interplay between scientific process and public policy is not new but has been a characteristic of most of the great turning points in human history. One need look no further than the dawning of the nuclear age to see that Fermi, Bohr, Oppenheimer, and Sakharov influenced today's world just as much as Roosevelt, Stalin, Churchill, Gandhi, and Hammarskjold did. It may be more important than ever for scientists to keep the doors of their laboratories open to political, economic, social, and ideological currents. The role of scientists as isolated explorers of the uncharted world of tomorrow must be reconciled with their role as committed, responsible citizens of the unsettled world of today.

The international agenda is varied and complex, but also promising. Advances are being made in a number of fields. The most notable of these may be that tensions between East and West are easing, brightening prospects for gains in peace and security and the settlement of regional conflicts. Should we not take advantage of this favorable atmosphere and direct our efforts toward the critical environment and development issues facing us? Many of these problems cannot be solved within the confines of the nation state, nor by maintaining the dichotomy between friend and foe. We must increase communication and exchange and cultivate greater pluralism and openness.

In 1987 the World Commission on Environment and Development, of which I was chairperson, presented its report, *Our Common Future*. The report sounded an urgent warning: Present trends cannot continue. They must be reversed. The Commission did not, however, add its voice to that of those who predict continuous negative trends and decline. The Commission envisioned a positive future. Never before in our history have we had so much knowledge, such sophisticated technology, and such wide access to resources. We have an opportunity to break the negative trends of the past. For this to happen, we need new concepts and new values based on a new global ethic. We must mobilize political will and human ingenuity. We need closer multilateral cooperation based on the recognition that nations are increasingly interdependent.

The World Commission offered the concept of sustainable development, which it defined as an approach toward meeting the needs and aspirations of present and future generations without compromising the ability of future generations to meet their needs. It is a concept that can mobilize broader political consensus, one on which the international community can and should build. It is a broad concept of social and economic progress. It requires political reform, access to knowledge and resources, and a more just and equitable distribution of wealth within and between nations. It demands that we move beyond compartmentalization and outmoded patterns to draw the very best of our intellectual and moral resources from every field of endeavor. Over the past couple of years, some progress has been made in the environmental field, both in terms of raising consciousness and in terms of taking on particular challenges, such as in the Montreal Protocol on the ozone layer and the Basel

Convention on hazardous wastes. The picture is very uneven, however, and the achievements do not justify complacency.

As far as development is concerned, the 1980s were a lost decade. Some countries did well, but in the Third World economic retrogression was widespread. Living standards have declined by one fifth in sub-Saharan Africa since 1970. Unsustainable, crushing burdens of debt and reverse financial flows, depressed commodity prices, protectionism, and abnormally high interest rates have all created an extremely unfavorable international climate for development in the Third World.

It is politically, economically, and morally unacceptable that there is a net transfer of resources from poor countries to rich ones. Nearly a billion people live in poverty and squalor, and the per capita income of some 50 developing countries has continued to decline over the past few years.

These trends must be reversed. As the World Commission pointed out, only growth can eliminate poverty. Only growth can create the capacity to solve environmental problems. But growth cannot be based on overexploitation of the resources of developing countries. It must be managed to enhance the resource base on which these countries all depend. We must create external conditions that will help rather than hinder developing countries in realizing their full potential.

We need a global consensus for economic growth in the 1990s. Such a unified plan must include the following attributes:

- Economic policy must be coordinated to promote vigorous, noninflationary economic growth. Major challenges include reducing the imbalance of payments between developing nations and the United States, Japan, and the Federal Republic of Germany, and making the surpluses of wealthy industrialized nations readily available to developing countries. The financial surpluses of the OECD (Organization for Economic Cooperation and Development) countries should be increasingly invested in developing countries rather than used to finance private consumption in the major industrialized countries.
- Policies should be adopted that will foster more stable exchange rates and increase access to markets on a global ba-

sis. Protectionism is a confrontational issue and a no-win game. Every year developing countries lose twice as much due to protectionism as the total they receive in development assistance. The benefits of free trade for both the North and the South are obvious.

- Policies should be devised that will sustain and improve commodity prices. Other policies should encourage and support diversification of the economies of the developing countries.
- Third World debt must be reduced. This problem requires major new efforts. For debt owed to multilateral institutions, the scheme based on a Nordic proposal to soften interest payments on such loans has been taken up by the World Bank. I believe this and similar schemes should be extended in the future. According to a very civilized, ancient legal provision, "If a man owes a debt, and the storm inundates his field and carries away the produce, or if the grain has not grown in the field, in that year he shall not make any return to the creditor, he shall alter his contract and he shall not pay interest for that year." This quote from the Code of Hammurabi, King of Babylon, dates from 2250 B.C. Four thousand years later the debt burdens, the environmental crisis, and the decline in the flows of resource transfers deserve equally civilized consideration.
- In addition to revamping our debt policies, we should increase development assistance for the poorer nations of the developing world, especially in Africa. In recent years, Norway has given around 1.1 percent of its gross national product in official development assistance to developing countries, and we are disappointed that the OECD average has declined to a meager 0.34 percent. Those donor countries that have been lagging behind should make renewed efforts in line with their abilities. The Soviet Union and Eastern Europe should also contribute to a far greater extent than they have so far. The developing countries have been declaring their readiness to do their part in terms of policy reforms and constructive negotiations.

A global consensus for economic growth in the 1990s must be consistent with sustainable development. It must take heed of ecological constraints. There are no sanctuaries on this planet. If 1990s are truly to be a decade during which we respond to the serious problems confronting the world, the issue of sustainable global development requires special, and urgent, attention.

The threats to the global environment have the potential to open our eyes and make us accept that North and South will have to forge an equal partnership. It is time to launch a new era of international cooperation. Issues like the debt crisis, trade policies, resources for the international financial institutions, harnessing technology for global benefit, strengthening the United Nations system, and specific major threats to the environment such as global warming are increasingly interrelated. Is it not appropriate to consider our economic and our environmental concerns together, given the critical links between the two?

Third World nations seem convinced that the poverty they endure is not a mere aberration of international economic relations that can be corrected by minor adjustments, but rather is the unspoken premise of the present economic order. Developing countries have had to produce more and sell more in order to earn money to service debt and pay for imports. The amount of coffee, cotton, or copper they must produce to buy a water pump, antibiotics, or a truck keeps increasing. This has caused people to place extra stress on the environment, which has fueled soil erosion, accelerated the cancerous process of desertification and deforestation, and begun to threaten the genetic diversity that is the basis for tomorrow's biotechnology, agriculture, and food supply.

Biotechnology warrants special consideration because of its potential effects on agriculture and food security in the Third World. It is difficult to imagine producing enough food to feed a doubled world population without employing the techniques and advances of biotechnology. The benefits of plant breeding and breeding of varieties with greater resistance and more rapid growth potential have been and will continue to be immense. But there are inherent dangers that could further widen the gap between poor and rich. These benefits may become available only to the rich, even though the genes employed in the process often originated in developing countries. Strong international corporations may dominate this field. Legal protection and very firm rules regarding rights of ownership may reduce the availability of products that are important for nutrition and the prevention of famine. Small-scale farmers in the Third World risk being victims in this process. If biotechnology produces substitutes for their crops, they may lose income and the ability to provide for their families.

These problems are avoidable. The industrialized countries have a responsibility for controlling market forces in this field and for promoting greater equity between developed and developing countries. The protection of intellectual property rights and royalties must be in a form that promotes research, provides for a fair distribution of financial benefits between inventors and the country of genetic origin, and, not least, makes the products of biotechnology available to those who need them.

The issues related to handling of biotechnology also pertain to our response to the possibility of global heating and climatic change. These events may require drastic changes in how we conduct our daily lives, and pose more severe threats to future development than any other challenge mankind has faced. A possible exception is the threat of nuclear war, but there is a decisive difference. At present nuclear war seems more remote than at any time since World War II, but unless we reduce our consumption of fossil fuels, we will be caught in the heat trap of global warming.

We may be about to alter the entire ecological balance of the earth. Plants and animals normally need hundreds of years to adjust to new climate conditions. Unless we make changes in our collective behavior, ecosystems will not have time to adjust. Deserts will spread. Crops will be lost. What will happen if we experience 2 years in succession in which summers are as dry as the one in 1988, or 10 such summers? What will happen to crop yields? Can we conceive of a doubling of food prices, or even food scarcity in the industrialized countries? The developed countries may be able to cope in the short run as long as they can pay for necessary imports, but that option will soon be lost to the developing countries. Can we imagine the effects on low-lying countries if sea level should rise according to predictions? How will we handle the political instability that will accompany increased migration as the number of environmental refugees continues to multiply?

These things may not happen, or may not be that drastic. But the potential risks are so high that we cannot sit back hoping that the problems will solve themselves.

The present generation has a great responsibility to control its use of limited resources, in particular the fossil fuels, which, when burned, contribute so significantly to the buildup of greenhouse gases in the atmosphere. We must recognize that the earth's atmosphere is a closed system. We do not get rid of our emissions when we vent them to the atmosphere. In fact, the current system is like a car that channels its fumes into the driver's compartment.

We must combat the myth that energy consumption can be allowed to grow unchecked. The industrialized countries have the greatest resources, both financially and technologically, to change production and consumption patterns. The developing countries will need much more energy in the future. Many of them have contributed only marginally to the greenhouse effect, and many of them will be most severely victimized by the effects of global warming. They must be allowed time to adapt, and a chance to increase their energy consumption and standards of living. It is quite clear that developing countries need assistance in order to avoid repeating the mistakes of industrialized countries. It is essential that energy-efficient technology be made available to developing countries even when they cannot pay market prices without assistance.

We need concerted international action. There are certain imperatives that must be vigorously pursued:

- We must agree on regional strategies for stabilizing and reducing emissions of greenhouse gases. Reforestation efforts must be included as a vital part of the carbon equation.

• We must strongly intensify our efforts to develop renewable forms of energy. Renewable energy should become the foundation of the global energy structure during the twenty-first century.

• We should speed up our efforts on international agreements to protect the atmosphere. There are different views on how to proceed on this issue. I urge that negotiations to limit emissions of greenhouse gases begin immediately.

Twenty-two heads of state or government took steps toward achieving these goals in March 1989, when they signed a declaration that set a standard for future achievements to protect the atmosphere. The Declaration of The Hague calls for more effective decision-making and enforcement mechanisms in international cooperation as well as greater solidarity among nations and between generations. The declaration calls for a new international authority with real powers. On occasion its power must be exercised even if unanimity cannot be reached. The principles we endorsed are radical, but a less ambitious approach will not serve us. The burden must be shared. That is why we called for fair and equitable assistance to compensate those developing countries that will be most severely affected by a changing climate but that have contributed only marginally to global warming.

In April 1989 the Norwegian government adopted a white paper in response to the proposals advanced in *Our Common Future*. The white paper is the Norwegian government's major policy document on sustainable development. It presents a plan that involves all ministries, not only that of environment, and implies change in attitudes and policies, and tough challenges for ministries such as energy, industry, transportation, finance, foreign affairs, and trade. The prime minister's office has been directly engaged in charting a course for the future that cuts across all of these sectors.

The issue of atmospheric pollution and climate change proved very difficult because Norway is fortunate in having vast hydropower resources. We do not burn coal or oil to produce electricity. Any reduction of carbon dioxide emissions in Norway would involve transportation. Nonetheless, Norway adopted a policy for stabilizing its carbon dioxide emissions during the 1990s, and by the year 2000 at the latest. By also reducing its emissions of CFCs and nitrogen oxides, Norway will be able to reduce its total emissions of greenhouse gases by the turn of the century. The government anticipates that further reductions will be possible thereafter. I believe we are the first country to make a political commitment to reduce carbon dioxide emissions, even though Norway contributes only 0.2 percent of the carbon dioxide emitted worldwide.

While even one nation can work to improve conditions in the global environment, ecological problems such as the ozone layer, global warming, and unsustainable use of the tropical forests clearly face mankind as a whole. To address these problems, additional resources will be needed. In the white paper, we propose, as a starting point, that industrialized countries allocate 0.1 percent of their gross domestic product to an "International Fund for the Atmosphere." Ideally, all countries would contribute. Much work is needed to make this proposal operational, and it will meet considerable resistance. But unless we establish a set of international support mechanisms, there is little chance that we will be able to reduce greenhouse emissions in time to slow the rate of global warming, or even give ourselves sufficient time to adapt.

To transform the essence of *Our Common Future* into reality will require broad participation. Every person can make a difference. Changes are the sum of individual actions based on common goals. A particular challenge goes to youth. More than ever before, we need a new generation—today's young people—that can use their energy and dedication to transform ideas into reality. Many of today's decision makers have yet to realize the peril in which the earth has been placed. I believe that *Our Common Future* can be an effective lever in the hands of youth and that its core concepts—that development must be sustainable, and that the environment and world economy are totally, permanently intertwined—transcend nationality, culture, ideology, and race. Young people will hold their governments responsible and accountable, and will build the foundation of their own future.

If we are earnest in our desire to solve some of the pressing problems facing humanity, we can no longer separate the global environment from political, economic, and moral issues. Environmental considerations must permeate all decisions, from consumer choices to national budgets to international agreements. We must learn that environmental considerations are part of unified management of our planet. This is our ethical challenge. This is our practical challenge. A challenge we must all accept.

APPENDIXES

A

Selected Reading

Ausubel, J. H., and H. E. Sladovich, editors. 1989. Technology and Environment. National Academy Press. Washington, D.C.

Clark, W. C., and R. E. Munn, editors. 1986. Sustainable Development of the Biosphere. Cambridge University Press. Cambridge.

DeFries, R., and T. Malone, editors. 1989. Global Change and Our Common Future: Papers from a Forum. National Academy Press. Washington, D.C.

NASA Earth System Sciences Committee. 1988. Earth System Science: A Closer View. Prepared for the NASA Advisory Council. (For availability, contact: Office for Interdisciplinary Earth Studies, University Corporation for Atmospheric Research, P.O. Box 3000, Boulder, CO 80307.)

Rosenberg, N. J., W. E. Easterling III, P. R. Crosson, and J. Darmstadter, editors. 1989. Greenhouse Warming: Abatement and Adaptation. Resources for the Future. Washington, D.C.

Schneider, S. H. 1989. Global Warming: Are We Entering the Greenhouse Century? Sierra Club Books. San Francisco.

Scientific American. 1989. Managing Planet Earth, Special Issue. New York. September.

Wilson, E. O., editor. 1988. Biodiversity. National Academy Press. Washington, D.C.

World Commission on Environment and Development. 1987. Our Common Future. Oxford University Press. Oxford.

World Resources Institute and International Institute for Environment and Development, in collaboration with United Nations Environment Programme. World Resources, 1988–1989. Basic Books. New York.

B

Forum on Global Change and Our Common Future: Agenda and Speakers

Tuesday, May 2, 1989

8:00 a.m. *Registration*

9:00 a.m. Welcome
THOMAS MALONE, St. Joseph College

9:10 a.m. Society's Stake in Global Change
WILLIAM RUCKELSHAUS, Browning-Ferris Industries

UNDERSTANDING GLOBAL CHANGE: THE SCIENCE

Chair: Thomas Malone, St. Joseph College

9:45 a.m. Historical Perspectives: Climate Changes Throughout the Millennia
JOHN KUTZBACH, University of Wisconsin

10:15 a.m. *Break*

10:25 a.m. Understanding Global Change: Earth as a System
FRANCIS BRETHERTON, University of Wisconsin

11:00 a.m. *Panel:* **The Earth System**
Moderator: DIGBY McLAREN, Royal Society of Canada

Atmosphere
MICHAEL McELROY, Harvard University

Oceans
JAMES McCARTHY, Harvard University

Terrestrial Ecosystems
PETER VITOUSEK, Stanford University

Human Dimensions
ROBERTA BALSTAD MILLER, National Science Foundation

12:45 p.m. *Lunch*

2:15 p.m. Human Causes of Global Change
B. L. TURNER II, Clark University

2:45 p.m. *Panel:* **Consequences**
Moderator: ROBERT McC. ADDAMS, Smithsonian Institution

Greenhouse Warming
JERRY MAHLMAN, Geophysical Fluid Dynamics Laboratory/National Oceanic and Atmospheric Administration

Stratospheric Ozone Depletion
SUSAN SOLOMON, Environmental Research Laboratory/National Oceanic and Atmospheric Administration

Break

Deforestation
ENEAS SALATI, Escola Superior de Agricultura, São Paulo, Brazil

Acid Deposition
DAVID SCHINDLER, Department of Fisheries and Oceans, Canada

Implications for Life on Earth
PAUL EHRLICH, Stanford University

5:30 p.m. *Recess*

7:30 p.m. Keynote Address and Franklin Lecture
Global Change and Our Common Future
GRO HARLEM BRUNDTLAND, Prime Minister of Norway

Wednesday, May 3, 1989

IMPACTS OF GLOBAL CHANGE

Chair: Robert Hoffmann, Smithsonian Institution

9:00 a.m. What Does Global Change Mean for Society?
LESTER BROWN, Worldwatch Institute

9:30 a.m. *Panel:* **Impacts**
Moderator: JOSE GOLDEMBERG, University of São Paulo, Brazil

Agriculture and Water Resources
THEODORE HULLAR, University of California, Davis

Break

Biodiversity
ROBERT PETERS, World Wildlife Fund

Sea Level
JAMES BROADUS, Woods Hole Oceanographic Institution

Industry
HUGH WYNNE-EDWARDS, Alcan, Canada

11:30 a.m. Implications of Global Change for Science Policy
ROBERT CORELL, National Science Foundation

12:00 p.m. *Lunch*

IMPLICATIONS FOR PUBLIC POLICY

Chair: Thomas Lovejoy, Smithsonian Institution

1:30 p.m. Options for Action
MARTIN HOLDGATE, International Union for Conservation of Nature and Natural Resources

2:00 p.m. View from the North
DIGBY McLAREN, Royal Society of Canada

2:30 p.m. View from the South
MARC DOUROJEANNI, The World Bank

3:00 p.m. *Break*

3:10 p.m. *Panel:* **Public Policy**
Moderator: JESSICA MATHEWS, World Resources Institute

THE HONORABLE CHARLES CACCIA, member of Parliament, Canada
THE HONORABLE JOHN CHAFEE, U.S. Senate
WILLIAM A. NITZE, U.S. Department of State
PAULO NOGUEIRA-NETO, University of São Paulo, Brazil
THE HONORABLE TIMOTHY WIRTH, U.S. Senate

5:30 p.m. *Recess*

Summary Panel

7:30 p.m. *Moderator:* THOMAS MALONE, St. Joseph College
ALAN HECHT, National Oceanic and Atmospheric Administration
RAFAEL HERRERA, Instituto Venezolano de Investigaciones Cientificas, Venezuela
JOHN P. HOLDREN, University of California, Berkeley
THOMAS LOVEJOY, Smithsonian Institution
STEPHEN H. SCHNEIDER, National Center for Atmospheric Research
ANNE WHYTE, International Development Research Centre, Canada

C

Committee on Global Change and Oversight Committee Members

COMMITTEE ON GLOBAL CHANGE
(U.S. National Committee for the IGBP)

HAROLD MOONEY, Stanford University, *Chairman*
PAUL G. RISSER, University of New Mexico, *Vice Chairman*
D. JAMES BAKER, Joint Oceanographic Institutions, Inc.
FRANCIS BRETHERTON, University of Wisconsin
KEVIN BURKE, Lunar and Planetary Institute
WILLIAM C. CLARK, Harvard University
MARGARET B. DAVIS, University of Minnesota
ROBERT E. DICKINSON, National Center for Atmospheric Research
JOHN IMBRIE, Brown University
ROBERT KATES, Brown University
THOMAS MALONE, St. Joseph College
MICHAEL B. McELROY, Harvard University
BERRIEN MOORE III, University of New Hampshire
ELLEN S. MOSLEY-THOMPSON, Ohio State University
PIERS J. SELLERS, University of Maryland

Ex-Officio Members
U.S. Members, ICSU Special Committee for the IGBP

JOHN A. EDDY, University Corporation for Atmospheric Research
JAMES J. McCARTHY, Harvard University
S. ICHTIAQUE RASOOL, National Aeronautics and Space Administration

Staff

JOHN S. PERRY, *Staff Director*
RUTH S. DeFRIES, *Staff Officer*

AD HOC COMMITTEE FOR OVERSIGHT OF THE COMMITTEE ON GLOBAL CHANGE

NORMAN HACKERMAN, Robert A. Welch Foundation (Chairman, Commission on Physical Sciences, Mathematics, and Resources), *Convenor*
ROBERT McC. ADAMS, Smithsonian Institution (Chairman, Commission on Behavioral and Social Sciences and Education)
BRUCE ALBERTS, University of California, San Francisco (Chairman, Commission on Life Sciences)
WILLIAM GORDON, Rice University (Foreign Secretary, National Academy of Sciences)

Staff

JOHN BURRIS, Executive Director, Commission on Life Sciences
VICTOR RABINOWITCH, Executive Director, Office of International Affairs
MYRON UMAN, Acting Executive Director, Commission on Physical Sciences, Mathematics, and Resources
SUZANNE WOOLSEY, Executive Director, Commission on Behavioral and Social Sciences and Education

D

Letter on Global Environmental Change to President-Elect George Bush

from the National Academy of Sciences, the National Academy of Engineering, and the Institute of Medicine

The Honorable George Bush
The President-Elect of the United States
Old Executive Office Building
Washington, D.C. 20501

Dear Mr. President-Elect:

The problem of global environmental change is now widely recognized as one of growing urgency that will require responses by your Administration. Our views on the problem and possible responses are attached.

Embedded in the diverse manifestations of this problem—global warming, ozone depletion, tropical deforestation, and acid deposition—are enormous challenges to science and engineering, to your Administration, and to the world community of nations. In many instances, data and analyses are incomplete and long-term effects remain indeterminate; in addition, there are costs to the economy embedded in any decisions made to address the problem. Yet, even with a continuing background of uncertainty, it is important to recognize that human activities are indeed changing the global environment. Prudent courses of action need to be initiated now to try to understand and predict these changes, and to move toward suitable policy responses.

As outlined in the paper that follows, we suggest three principal actions that your Administration can take against the possibility that we may be conducting an irreversible experiment with our planet.

First, the issue of global environmental change must be made more prominent in the scientific, political, and foreign policy agendas of the United States. While the issue has been given more emphasis in the past several months, the need now is for sustained and systematic attention to economic and environmental policies bearing on global change.

Second, a focal point for the deployment of national resources to address the issue needs to be established to coordinate the diverse activities related to the problem. Such activities under way within the government are often poorly coordinated and insufficiently focused on the necessary linkages between scientific understanding and policy options. Further, policies must be such that they enhance the resiliency, robustness, and range of options of the affected sectors, from energy production to agriculture to coastal development.

Third, specific actions should be taken by your Administration to address those aspects of the problem that are reasonably well understood. Thus, we suggest that you consider the following findings and responses:

- Both global warming and acid deposition are linked to heavy reliance on fossil fuels. Therefore, efforts to enhance both energy efficiency and conservation should be strengthened; means for increasing the use of the "cleanest" fossil fuels, such as natural gas, should be considered, but in the context of appraising their finite availability; and alternative non-fossil fuel energy sources, such as nuclear reactors and solar energy, need to be reappraised with more emphasis put on their use in a safe and publicly acceptable manner.

- The Montreal Protocol is an important first step toward controlling stratopheric ozone depletion. However, provisions in the Protocol should be called upon to encourage deeper reductions in emissions of ozone-destroying chemicals with a total phaseout of chlorofluorocarbon products on a reasonable time scale.
- Global warming will affect sea levels, agricultural systems, forestry, and water resources. It is not too soon to assess possible impacts and responses.
- Tropical forests are being cut down at an alarming rate. Mitigating strategies that consider the economic and other forces driving this destruction need to be implemented in collaboration with the affected nations.
- Improved predictions of the future course of the global environment require new and innovative approaches to studying the earth and its environment. A substantial investment in research and ground- and space-based monitoring activities is needed.

This is a broad agenda, and a challenging one. However, we believe that it offers a sensible approach to problems that urgently require presidential leadership and whose consequences, if realized, could be severe for all nations. Leadership by the United States could be a powerful force for uniting the nations of the world in a common endeavor vitally important to all.

Yours sincerely,

FRANK PRESS
President
National Academy of Sciences

ROBERT M. WHITE
President
National Academy of Engineering

SAMUEL O. THIER
President
Institute of Medicine

PREFACE

Early in 1988, the Presidents of the National Academies of Science and Engineering and the Institute of Medicine, with the concurrence of their Councils, decided on the preparation of a small number of "White Papers" for the incoming Administration. The intent was to summarize concisely a few critical national and global issues to which science and technology were central, and to suggest presidential options on these issues.

Global environmental change emerged at once as a leading candidate for such a paper. Concern about man-made changes in the atmosphere and consequent changes in climate and the ozone shield has been mounting for more than a decade. A broader range of issues, including tropical deforestation and regional degradation in water and air quality, has heightened public awareness. A major international research program to build an integrated base of knowledge to address these problems is taking shape. Thus, an early decision was made to develop a paper on global environmental change, along with papers on presidential science advising, the AIDS problem, and national space policy.

Topical outlines of the principal scientific problems and elements of the current scientific consensus were prepared with the aid of the staff and members of various National Research Council groups. These outlines were broadly circulated for comment. A revised outline, together with an exploratory draft, provided the basis for a searching discussion by a group of invited senior scientists* at an ad hoc meeting held in Boulder, Colorado, on

*Participants in the ad hoc meeting were C. J. Pings, University of Southern California (*Chairman*); James Anderson, Harvard University; Robert Fri, Resources for the Future; Stephen H. Schneider, National Center for Atmospheric Research; Thomas F. Malone, St. Joseph College; William Nordhaus, Yale University; Arthur Johnson, University of Pennsylvania; Robert McC. Adams, Smithsonian Institution; Jerry Melillo,

August 15-16, 1988. The views elicited in this discussion played a major role in the subsequent development of the paper by the Council groups.

The development of the paper was coordinated by John S. Perry and Ruth S. DeFries of the staff of the NRC Committee on Global Change and by Norman Metzger of the NRC Executive Office.

Woods Hole Oceanographic Institution; and Sir Crispin Tickell, United Kingdom Ambassador to the United Nations.

GLOBAL ENVIRONMENTAL CHANGE

Human activities now match or even surpass nature as an agent of change in the global environment, as evidenced by a growing list of seemingly diverse human-induced environmental changes that have gripped public attention in recent years:

- rapid changes in the global atmosphere due to fossil fuel combustion and industrial activities predicted to change global climate;
- massive ozone depletion over the Antarctic and lesser decreases over the rest of the globe, both attributed to emissions of chlorofluorocarbons;
- large-scale destruction of tropical forests for timber, fuel, conversion to agriculture, and economic development, with consequent additions to the "greenhouse" effect and losses in plant and animal species;
- damaged lakes in New England, Canada, and Scandinavia associated with acid deposition from fossil fuel combustion.

Although the full long-term implications of these changes are as yet unclear, there is a growing perception that the future welfare of human society is to an unknown degree at risk. *Our current scientific understanding amply justifies these concerns, but also presents opportunities for effective presidential leadership and action.*

Recent events have illustrated the possible social, economic, and political implications of future global environmental change. Widespread droughts in the early 1970s set the stage for major worldwide fluctuations in grain prices. Damage to lakes and forests has created major political tensions in North America and Europe. Continued droughts in Africa have stressed the aid and relief machinery of the developed nations of the north.

The individual environmental problems that have come to public attention are intricately and inescapably interlinked, both scientifically and politically. Scientifically, their resolution requires an understanding of the physical, chemical, and biological processes that govern the earth, and of the interaction of these processes in the entire earth system. Politically, policy options to address these problems converge on the need for internationally accepted actions relating to energy, technology, land use, and economic development. Their implementation will require U.S. commitment and the development of improved mechanisms for international collaboration.

Such issues challenge political leadership. The consequences of global changes to individual political constituencies are probably several decades away, unclear in nature, but potentially serious. Remedial and adaptive actions are likely to be expensive in the short term, whereas their full benefits may be enjoyed only by future generations. Many possible courses of action in policies relating to research, energy, development, and international relations involve multiple and interlinked costs and benefits transcending both national and generational bounds. Such problems can only be assessed, and prudent long-term response policies can only be effectively developed, at the highest political levels. Global change is quintessentially an issue for leadership at the level of heads of state. The United States, in your presidential term, is well positioned to play a world leadership role.

WHAT ACTIONS ARE SUGGESTED?

It is now timely to consider prudent policy actions in three areas: *prevention*, *adaptation*, and *research*. These actions, which can only be set in motion by presidential leadership, are discussed below, followed by synopses of the current scientific understanding of the components of global environmental change of current concern.

Preventing Global Environmental Change

We are already irrevocably committed to major global change in the years ahead. The elevated concentrations of greenhouse gases produced to date by human activities will persist for many centuries and will slowly change the climate of the earth, regardless of our actions. The chlorofluorocarbons (CFCs) that are depleting the ozone shield have lifetimes on the order of a century. Complex tropical forest ecosystems have the ability to regenerate, but this depends on the way they are managed, and regeneration will occur slowly at best.

While global environmental change cannot be stopped, the pace of change can be slowed. Put another way, we cannot buy absolute security against environmental risk, however much we are willing to pay; but we may be able to reduce environmental damage and risk markedly by prudent policy actions outlined below.

Energy Policy

Production of energy from fossil fuels is the root of several of the agents of global environmental change, notably climatic change and acid deposition. There has been great concern, and considerable remedial action, with respect to emissions of sulfur and nitrogen compounds from fossil fuel combustion. Moreover, reductions in fossil fuel combustion address the continuing concerns for regional air quality and acid deposition, the growing concerns for global climate, and our growing dependency on foreign energy resources, i.e., petroleum.

To prepare for the possibility that the use of fossil fuels (particularly coal) may need to be reduced in the years ahead, *we should explore means for reducing energy demand without impeding economic growth, for example, by using fossil fuels with greater efficiency, as well as incentives and other means to increase use of the "cleanest" fossil fuels such as natural gas.* However, the finite availability of natural gas, a premium energy source and valuable chemical feedstock, also needs to be considered. The potential for energy efficiency gains and fuel-switching is demonstrated by the responses to the energy price shocks of the early 1970s.

A vigorous research and development program should give increased emphasis to energy sources that do not produce carbon dioxide. New technologies for publicly acceptable nuclear reactors—so-called "safe reactors"—should receive attention in an R&D program on alternative energy sources. Geothermal, wind, solar, and other renewable energy sources warrant reexamination to see if recent progress has improved their economic feasibility as possible substitutes for fossil fuels in the years ahead.

Actions by the United States alone, while setting a good example, would in themselves be only marginally effective in slowing the rate of global carbon dioxide emissions. In the decades to come, major demands for energy will come from the developing nations. These nations will have to evaluate the needs of their own economic progress together with protection of the global environment. Thus, energy policy issues are likely to be on the international political agenda.

Policies to Reduce Industrial Emissions

The Montreal Protocol to limit emissions of CFCs, signed in the fall of 1987 by 31 nations—including the United States—is an unprecedented example of international cooperation to prevent global environmental deterioration. *The provisions of the Montreal Protocol could be called on to urge deeper cuts in the production of CFCs, to accelerate the timetable for their reduction, and to urge all countries to sign and enforce the Protocol.*

Technology exists to eliminate most of the industrial emissions that cause acid deposition. The use of clean, low-sulfur coal and combustion and cleaning technologies can significantly reduce emissions in power plants. However, some emission-control techniques may be energy-intensive, thus increasing carbon dioxide emissions, and may produce solid and liquid waste products that are difficult to deal with. Thus, policies to control acid deposition involve trade-offs with measures to address other environmental problems, and often require international coordination. In any case, there are political and economic components in any decision on dealing with acid deposition; and to date such decisions have been deferred by the past two Administrations in favor of more research. We believe that the sources of acid deposition, the technology to limit emissions, and the associated costs and political risks are now sufficiently understood that further deferral in favor of more analysis is unwarranted.

Development Policy

The future course of a number of the global environmental issues discussed above will be strongly influenced by the manner in which the economies of the less developed nations evolve over the next several decades. A nation's impact on its own environment, and on that of the globe, clearly depends on its population level, standard of living, and the technologies used to support that standard. Uncontrolled growth of populations in developing countries seeking better standards of living with inefficient technologies can only lead to continued damage to the local and global environments. *Development aid policies should focus on improving the indigenous scientific and technological knowledge base and infrastructure of the developing countries as a foundation for environmentally sustainable economic development to address these nations' aspirations for a better life. Development aid decisions should include a consideration of environmental impacts.*

Adapting to Global Environmental Change

If we do no more than slow the pace of environmental change, thus gaining time to deal with its effects, we will have achieved much. Policy options to enhance our ability to cope with environmental change should therefore also have high priority. Because the detailed evolution of the expected changes cannot at present be predicted, *policies should seek to enhance the resilience, robustness, and range of options of the sectors likely to be affected.* For example, the likelihood of rising sea levels should clearly be taken explicitly into account in the planning of coastal development, construction of port facilities, location of waste disposal dumps, and so on. Possible changes in the frequency and severity of extreme events—floods, droughts, and severe storms—should also be considered. The possibilities for significant changes in regional hydrology should be considered in planning for major water-resource projects. Better understanding of the response of crops and natural vegetation to changes in climate and atmospheric composition is needed. The already impressive adaptability of agriculture may be further enhanced by investments in the development of improved crop strains, the exploitation of new crops, and technology for improving water utilization.

Options for adaptation are numerous for rich, technologically advanced nations such as ours, but fewer for poor nations. It will be in the interests of all to assist the poorer nations in reducing their vulnerability to environmental change by developing an effective range of adaptations.

Improving Understanding of Global Environmental Change

We still lack sufficient scientific understanding to predict confidently the detailed evolution of global and regional change in the environment. Obtaining this knowledge will require *strong support for research already under way, nationally and internationally, and support for new long-term programs of interdisciplinary, international research.* Monitoring global change will involve coordinated, long-term observations, both from space and on the ground. Long-term commitments must be made for resources that will make uninterrupted observations possible over the decades to come.

The acquisition of knowledge through research and monitoring is expensive, although not nearly as costly as the possible consequences of ignorance. We are already investing large resources in the study of the earth, but these efforts require augmentation and coordination.

Institutional Needs

However, current institutions and mechanisms in government and in the scientific community—both in our country and in the international community—need strengthening to assure a strong scientific program for understanding the global environment and developing timely, well-coordinated, and effective policies. *The existing mechanisms for coordination between government agencies are not adequate to address the complex scientific and policy implications of global change.*

Mechanisms exist within the federal government that could play a major role in directing the massive and diverse resources of government on this problem; notably, the interagency Committee on Earth Sciences of the Federal Coordinating Council on Science, Engineering, and Technology. However, effective direction of these governmental resources cannot be achieved solely by the normal processes of interagency coordination. Even more complex issues are posed by the need for collaboration between sovereign nations. Effective leadership and direction in this area is clearly required. The issue of global environmental change must have a prominent place on the scientific, political, and foreign policy agendas of the United States.

PRINCIPAL SCIENTIFIC ISSUES

The policy recommendations suggested above are shaped by our current understanding of the various aspects of global environmental change, particularly by the considerable uncertainties in our knowledge. Reducing these uncertainties would lead to more informed and effective policy decisions.

Climate Change

A central issue in global environmental change is the effect of changing atmospheric composition on global climate. The greenhouse effect—the trapping of the sun's heat near the earth's surface by small concentrations of certain gases in the atmosphere—has been well understood since the eighteenth century and is arguably the most firmly established principle in atmospheric science. These "greenhouse gases" include water vapor, carbon dioxide, methane, and the CFCs.

In the nineteenth century, marked growth in carbon dioxide concentrations began—from about 280 parts per million (ppm) in 1850 to 350 ppm today. This increase can be explained by the burning of vast quantities of fossil fuels, the massive conversion of carbon-rich forests to farmland and pasture, and the absorption of about half of the emitted carbon dioxide in the ocean. Today, fossil fuel use injects about 5 billion tons of carbon into the atmosphere each year, and deforestation adds at least 1 billion tons more. As a result, carbon dioxide concentrations are continuing to increase at about 0.4 percent per year.

Other greenhouse gases are also increasing. Methane, emitted from oil and gas wells, rice paddies, bogs, tundra, and increasing populations of creatures ranging from cows to termites, is increasing at over 1 percent per year. The principal CFCs, despite recent limitations, are still increasing at several percent per year. Other gases such as nitrous oxide and tropospheric ozone are similarly increasing. The combined effect of each year's increase in these gases is roughly equal to that of the carbon dioxide increase alone.

Estimating the influence of continuing increases in greenhouse gases on future climate requires quantitative models of the climate system, as well as predictions of how human actions will influence future emissions. Based on our current understanding of natural processes and plausible projections of population growth, technology, and economic development, it is expected that before the middle of the next century the combined effect of increases in all greenhouse gases will produce an impact on the climate roughly equivalent to a doubling of preindustrial levels of carbon dioxide alone. Numerical models of the climate system yield estimates of the effect on long-term global mean temperatures that range between 1 and 5°C (2 to 9°F). The rate of climate change implied by these estimates is as much as 100 times greater than the average rate of change since the last glacial period (when global temperatures were about 5°C cooler than today and ice sheets covered our northern tier of states), and about 10 times greater than the fastest known sustained climate change during the last 10,000 years.

Recent observations of increased temperatures are consistent with predicted changes, but we do not yet know if they actually confirm that greenhouse warming is indeed in progress. Individual disastrous climatic events, such as a dry summer or a vigorous hurricane season, cannot be attributed to greenhouse warming.

The expected rapid rate of climate change greatly complicates the task of predicting the changes most important to human society—the timing and regional distribution of changes, and the frequency and severity of extreme events. Warming should lead to worldwide changes in many features of the hydrological cycle, with consequences for soil moisture, river flow, glacier extent, and the distribution of crop zones. In addition, there are reasons to expect that warmer ocean temperatures will lead to more violent tropical storms.

The detailed consequences associated with this expected warming are still unclear. Complex changes in temperature, precipitation, soil moisture, and storm patterns could result in a web of economic impacts.

- Climatic zones and storm tracks may be expected to shift poleward. Crop zones and natural ecosystems can be expected to migrate with the changing climate, although the extent of movement is uncertain. Major shifts could clearly have far-reaching economic, social, and political consequences.
- Melting of land-borne glaciers and thermal expansion of sea water are expected to raise global sea levels significantly over the next century. Projections of the amount of rise range from tens of centimeters to as much as 3 meters, with reasonable estimates centering on a 1 meter rise. Rising sea levels will increase the already troublesome rates of coastal erosion and loss of wetlands, while

increased saltwater intrusion would impair water supplies and agriculture in coastal areas.

- Warming is expected to be considerably greater in polar latitudes than in the tropics, and sea ice should diminish. A navigable Arctic Ocean would have major national security implications.
- Changes in rainfall patterns are likely, and some studies indicate greater summer dryness in midlatitude continental regions. Regional changes in water supply and quality may have significant economic and social consequences.
- Adverse climate changes may be difficult to accommodate in developing countries, where resources are not available to adapt to changing conditions. Growing pressures for migration may result.
- Implications for human health may result from changes in the range of disease vectors (e.g., insects and rodents) and in the frequency and intensity of extreme weather known to influence mortality rates.

There is a disturbing possibility for surprises as the climate system rapidly changes. The record of the past, notably as revealed by ice core records, shows that the climate system can exhibit large and rapid changes in response to slow changes in natural forcing. Today, humanity is imposing new stresses on the system, raising the question of possibly triggering climatic surprises in our own times.

Tropical Deforestation

Tropical forests are being rapidly destroyed because of the pressure of growing local populations for agricultural land and fuel wood supplies, and the strong world markets for tropical hardwoods and animal products from tropical pastures. Satellite data provide reasonably reliable estimates of the rates of deforestation; if these rates continue, the planet will be virtually denuded of tropical forests early in the next century.

This massive change in the earth's vegetation has many

consequences. As noted above, the associated carbon dioxide release adds about 20 percent to the release from fossil fuel combustion. Other troublesome gases such as nitrogen oxides and ozone are also produced directly or indirectly from burning to clear land. Deforestation can also lead to long-term degradation of soil fertility, and is intricately linked to significant changes in regional hydrology and climate. Perhaps the most serious long-term impact of tropical deforestation is the loss of plant and animal species. This loss of genetic resources may rob us of valuable options for future resources for food, industry, and health. Many of these changes are irreversible, and the full range of their long-term consequences is poorly understood.

Stratospheric Ozone Depletion

Ozone is naturally produced in the stratosphere by the interaction of sunlight with oxygen. Although present only in small concentrations, it blocks out highly damaging solar ultraviolet radiation, and is thus essential to life.

Observed ozone concentrations (a few parts per million) are maintained by a balance between ozone production and natural ozone-destroying chain reactions involving extremely small concentrations of nitrogen, hydrogen, and chlorine compounds. The CFCs, synthetic chemicals widely used in industry and consumer products, are known to enter the stratosphere, where they provide a new primary source of chlorine. This relationship between surface release of CFCs and global stratospheric ozone loss was identified in 1974.

The discovery of the Antarctic ozone "hole"—the abnormal depletion of ozone over the entire Antarctic continent in the early Southern Hemisphere spring—was brought to world attention in 1985. In the last 2 years, ground-based and high-altitude aircraft experiments have indicated that the ozone depletion is related to CFC concentrations, and laboratory experiments are clarifying our understanding of the mechanisms involved.

Small worldwide decreases in total ozone have recently been detected through careful analysis of surface-based and satellite observations. These changes amount to a few percent, and vary markedly by latitude and season. Such systematic long-term declines in ozone are accompanied by large natural seasonal and interannual variability, which can mask underlying trends.

The best understood implications of stratospheric ozone depletion lie in human health. Changes in total ozone of 5 percent in the middle latitudes would raise ultraviolet radiation levels at the surface enough to increase skin cancer incidence by more than 10 percent. Immunological changes also may result from ultraviolet exposure. Effects on plants and animals are poorly understood, although there are grounds for concern. Effects on food chains in the world ocean, with ultimate effects on fish harvests, as well as effects on agricultural crops and natural systems are also likely but have not yet been assessed.

Acid Deposition

Fossil fuel combustion in power plants and in autos produces oxides of sulfur and nitrogen, which are converted to acids in the atmosphere. These compounds are transported in the atmosphere over long distances and are responsible for reduced visibility and increased deposition of acids on the ground across large regions.

Increased deposition of acids in forested areas of the northeastern United States, Canada, and northern Europe that have low-alkalinity waters and thin, acid soils has resulted in the acidification of surface waters, sometimes to the extent that they no longer can support normal fish populations. The linkage between fossil fuel combustion and acidification has been demonstrated.

Acid deposition may be considered as one aspect of a changing "global chemical climate," with many varied and interlinked effects. For example, changes in the "chemical climate" have accelerated the corrosion and erosion of buildings and historic monuments. A serious concern is the possibility that changes in the deposition of nitrogen and sulfur compounds have significant effects on forest productivity. However, the full impact of acid deposition and other pollutants on forest productivity and its relationship to the decline of some North American, Canadian, and European forest species is not fully understood, and the mechanisms of forest damage are not clearly established.

Another aspect of the changing chemical climate is increased concentrations of ozone at ground level, resulting from reactions between nitrogen oxides, hydrocarbons, and sunlight. Ground-level ozone increase has also been implicated in human health problems and significant crop losses. Although the degree of increase is uncertain, ambient levels of ground-level ozone in the summertime have been associated in experimental trials with a loss of growth and altered physiological function of forest trees.

Finally, ozone in the lower atmosphere acts as a greenhouse gas, and the effects of a changing chemical climate on biological processes undoubtedly influence the fluxes of other greenhouse gases as well.

As a final comment, we believe that global environmental change may well be the most pressing international issue of the next century. The United States is well positioned to play a leadership role in coping with and gaining an international consensus on this difficult issue. Indeed, the United States in the past has had a major role in examining the problem—whether through the carbon dioxide measurements done on Mauna Loa, the analysis and development of the experimental evidence for the relationship of the Antarctic ozone depletion and CFCs, the substantial contributions of American scientists to creating mathematical models of climate, or United States leadership on the Montreal Protocol.

Index